Total Manufacturing Solutions

How to stay ahead of competition and management fashions by customizing total manufacturing success factors

Ron Basu
and
J. Nevan Wright

Butterworth-Heinemann
Linacre House, Jordan Hill, Oxford OX2 8DP
225 Wildwood Avenue, Woburn, MA 01801-2041
A division of Reed Educational and Professional Publishing Ltd

A member of the Reed Elsevier plc group

OXFORD   BOSTON   JOHANNESBURG
MELBOURNE   NEW DELHI   SINGAPORE

First published 1997
Paperback edition 1998

© Ron Basu and J. Nevan Wright 1997

All rights reserved. No part of this publication may be reproduced in any material form (including photocopying or storing in any medium by electronic means and whether or not transiently or incidentally to some other use of this publication) without the written permission of the copyright holder except in accordance with the provisions of the Copyright, Designs and Patents Act 1988 or under the terms of a licence issued by the Copyright Licensing Agency Ltd, 90 Tottenham Court Road, London, England W1P 9HE. Applications for the copyright holder's written permission to reproduce any part of this publication should be addressed to the publishers

British Library Cataloguing in Publication Data
A catalogue record for this book is available from the British Library

ISBN 0 7506 4041 3

Typeset by Avocet Typeset, Brill, Aylesbury, Bucks
Printed and bound in Great Britain by
Biddles Ltd, Guildford and King's Lynn

Contents

Foreword — vii
Preface — ix
Acknowledgements — xiii

1. Introduction: why total manufacturing solutions? — 1
2. Understanding total manufacturing solutions — 9
3. Marketing and innovation — 17
4. Supply-chain management — 33
5. Environment and safety — 57
6. Manufacturing facilities — 70
7. Procedures — 109
8. People — 143
9. 200 questions — 171
10. Data recording and analysis — 189
11. Mission statements — 205
12. Gap analysis — 215
13. Improvement strategy — 227
14. Implementation plan — 235
15. Final analysis: 'the big picture' — 252

Index — 263

Foreword

As Bertrand Russell once said, 'even when the experts all agree they may well be mistaken'. But one thing that cannot be doubted is that real wealth comes from the physical transformation of inputs into outputs. It would be a mistake to doubt this. If a country does not have a strong manufacturing base then it won't be able adequately to support service industries. This book recognizes this and is brave enough to point out the dangers of overlooking this fact by, for example, relying on the Third World to do the manual work to produce the goods while we provide the services.

The authors, Ron Basu and Nevan Wright, have not taken a populist approach. Rather, they have chosen to go behind the popular gloss and to get back to basics. Their philosophy is that if the product is simply not up to standard, it really doesn't matter how good the service, or what latest management fad the corporation has adopted. Nor does it matter how well the mission statement reads. Consumers want products not poetry.

The book gives a new approach to getting back to manufacturing basics. It offers a practical method, based on self analysis, to create an efficient, value-added manufacturing organization. Quality and customer focus are not ignored and indeed the authors provide a new insight into how to create a quality culture. But the central theme is that if the transformation process is not efficient, if value is not being added, then no amount of service is going to matter - for in today's marketplace customers expect first and foremost quality products.

It is interesting to note that Ron Basu was born in India, although he has lived most of his life in Britain, and that Nevan Wright is a New Zealander who spends some time each year in England at the Henley Management College. Perhaps it is their internationalism which allows them to stand back and dispassionately analyse what needs correction in our manufacturing industries.

This is not to suggest that the book is written in a critical way; far from it. Their comments are reasoned, and the solutions are drawn from the practical manufacturing experience of both of them. Between them they blend their experiences in academia, consultancy and industry to bring a refreshing down-to-earth approach and to give practical manufacturing solutions. This is a practical book for practical people.

It would be wrong to think of this as just another book on quality. Most of us surely are getting a little tired of new books aiming to popularize or exploit probably transient fads. After all, what else can be written on the subject? Providing we realize that quality does give the competitive edge, and that quality is the concern of all involved in the organization, then surely that is all that matters.

This therefore is a book for those who are not satisfied with the theory, and who want to know how to achieve things in practice. It will show you not only why you cannot afford not to get started, but how to do so - and help you achieve tangible benefits.

Ray Wild
Henley Management College
September 1996

Preface

> *It is ignoble to say one thing and to mean another:*
> *How much more so to write one thing and to mean another.*
> Seneca

Our work on this book started with an extraordinary coincidence. About a year ago both of us were discussing, almost at the same time, with Ray Wild, at the Henley Management College, an outline for a practical book on manufacturing. Ray identified our common theme and duly arranged for us to meet, and the result is *Total Manufacturing Solutions*. For getting us together and making it happen we are both indebted to Ray.

In our first discussion we both agreed, based on our combined experience as practising managers and from working in academia, that although in recent years many books had appeared on the subjects of excellence, quality and management, there did appear to be a lack of practical books which get back to basics, show the reader how to get started, and show how to make things actually happen. We both agreed that before people can get started, or make things happen, that they need to know where they are. In other words you can't improve if you don't know where you need to improve. We agreed that some form of benchmarking was needed so that a company could determine how they stacked up against international, or world class standards.

We believe there is a shift of focus to manufacturing as the competitive weapon in the global market. This shift cannot be faulted. What can be faulted is the host of management fads and three-letter acronyms (often 'solutions in search of problems') which are not capable of delivering the desired results. In addition many companies achieve manufacturing excellence in factories, but due to a failure to address the whole spectrum of the supply chain, they are not achieving a corresponding improvement in business performance. There are also many managers disillusioned with 'benchmarking' because of poor performances by external consultants.

To address these challenges we have developed a systematic approach to total manufacturing solutions.

We recognize that all organizations are different, and what will work for one organization may not necessarily be appropriate for another. Our book shows how manufacturers can self-examine their own organization and determine strengths and weaknesses. The approach is fundamental and simple. Total manufacturing comprises six pillars underpinned by 20 foundation stones or defined areas. We provide 200 questions designed for a company to carry out its own benchmarking. We show from this self-examination which improvement strategy can best be adopted to enable an organization to become a world class enterprise. Many books discuss what might be done, we actually go further and detail practical ways of how to make things happen.

Our philosophy is that it is both the achievement of manufacturing performance and the application of best practices that give a company a sustainable leading edge. Moreover, we believe that history proves that strong nations are those that are strong in manufacturing. The Industrial Revolution made Britain Great, mass production and innovation made the United States a world power, and excellence in manufacturing performance has made Japan the strong industrial nation it is today. To sustain a healthy national economy a country needs efficient world class manufacturers. Today's issues are globalization and international competition. No longer are companies protected by national barriers and tariffs. For some this may be perceived as a threat, but conversely, a leading edge organization will see this as an opportunity. This book shows how to secure a leading position and how to make the most of emerging global opportunities through implementing customized change programmes.

In our professional experience we have learnt that each organization has something unique to offer. In addition we recognize that most companies, in spite of the 'flattening' of their organization structure, are still well endowed with experienced executives who are most qualified to evaluate their own business needs. However, the missing link is a comprehensive outward looking approach. This book fills this missing link.

Total manufacturing permeates every function in a manufacturing organization, so that an understanding of all of the 'pillars' requires the consideration of many specialized topics. We have therefore included chapters on the basic concepts of business functions including marketing, innovation, supply chain, environmental care, quality systems, financial management, information technology and human resources. We have tried to treat all of these subjects in sufficient detail, but often at a basic level, to reach a broad cross-section of business managers. Some of these chapters may stand alone, but the importance of their interaction and the underlying theme can only be appreciated by reading the book from the beginning to the end.

In spite of all our efforts, there are still likely to be gaps in the book, partly by our oversight and party because we are venturing onto a new path of 'self-analysis'.

In our quest for never ending improvement, our efforts are now being channelled into collecting and writing case studies where manufacturing excellence has been achieved, and, in developing a software package to assist with the benchmarking process based upon the 200 questions and the methodology given in this book.

The authors would like to point out that the opinions expressed in this book are their own and not necessarily those of their employers.

We believe that what has resulted is a book which people will read and reread, and which they will pass onto friends and colleagues. This book is manufacturing focused and the reason for this is given in Chapter 1, but the messages are equally applicable to service industries. Our book is aimed at a broad cross-section of readership including:

- Senior executives (practitioners no matter what their function) will find that *Total Manufacturing Solutions* will give them a methodology to achieve world class manufacturing status for their organization. They will find our method of self-assessment and internal and external benchmarking relevant and practical.
- Professional management consultants will find the comprehensive approach of this book a useful guide for benchmarking studies and for training seminars.
- Business travellers and managers and staff of all disciplines will read this book to enhance their knowledge base. They will be able to apply the techniques described without disruption to their own working environment.
- Management schools, such as the Henley Management College, are likely to use this book as a reference. This will also apply to universities and other tertiary institutions.

Finally as Thomas Edison once said, 'Your idea has to be original only in its adaptation to the problem you're currently working on'. We hope that our book is a blueprint for manufacturing innovation and a breakthrough in performance improvement.

Acknowledgements

I am indebted to Professor Ray Wild for encouraging the inception of this book and for kindly agreeing to write the Foreword.

The ideas that I am sharing with the readers of this book have been shaped by many colleagues, clients and companies across the world over several years. I am also grateful to John Russell Associates Limited for providing details of their Inter Company Productivity Group Research and to Dr Michael Cross for making available his work on self test questions.

I wish to extend my warmest regards to my co-author Nevan Wright and his wife Joy for their invaluable input and humour during this project.

Jacquie Shanahan and Caroline Struthers, our editors at Butterworth-Heinemann, are to be congratulated for making the publishing process look deceptively easy. Special thanks go to Lilibeth Sakate for all her hard work with graphics and research.

Finally, for my part I regard this book as a family rather than an individual enterprise, since without the support and forbearance of my wife Moira and my children Bonnie and Robi during my self-imposed immersion in the book I might have believed that life consisted entirely of seeking manufacturing solutions.

Ron Basu

Like Ron, I too wish to acknowledge the support of Professor Ray Wild and additionally from my colleagues at the Henley Management College. I also wish to acknowledge the great input of Ron into this book. Ron proved to be a most agreeable person to work with.

A book such as this is based on my own experiences, the experience of others and from a wide reading of contemporary management issues. It would not be possible to acknowledge the source of all the ideas that have surfaced in this book. In some cases the source has been lost in the mists of time, on other occasions people have fed me back ideas or suggestions that I am sure were first propounded by me in past lectures or articles! Suffice it to say that Ron and I have made every endeavour to credit materials of others in the list of references at the end of each chapter. If there are any omissions, they are sincerely regretted.

Of the many people I have worked with over the years, there is one man who stands out more that any other, who in my opinion was outstanding in his efforts to gain manufacturing excellence for the several companies he managed. That man is Graeme Fincher, whose credo is to keep it simple, to make it happen, to have pride in a quality product and to respect, involve and trust people who perform. Invariably, fuelled by his example, Graeme's people, myself included, did perform.

Finally I dedicate this book to my soulmate Joy, who put in long unpaid hours typing and correcting for Ron and I. She was always quick to bring me back to earth if my prose became too esoteric.

J. Nevan Wright

1

Introduction: why total manufacturing solutions?

> *There is a tide in the affairs of men,*
> *Which, taken at the flood, leads on to fortune;*
> *Omitted, all the voyage of their life*
> *Is bound in shallows and in miseries.*
> *On such a full sea are we now afloat,*
> *And we must take the current when it serves,*
> *Or lose our ventures.*
>
> William Shakespeare

In the 1980s service rather than manufacturing was seen by many in the West as the way of the future for business. In the 1990s the inherent weaknesses in the reliance on service are being increasingly exposed. It is now being recognized that in a global marketplace manufacturing is the main competitive weapon. Real wealth can only come from the physical adding of value in the manufacture of tangible products.

To some of us the move away from manufacturing and the reliance on the Third World to do the work and to produce the goods (while we in the West provided the services of banking, insurance, accounting, and consultancy), were always seen as recipes for disaster. History shows that it is the manufacturing countries, those that are actually making something and physically adding value, who will become strong, self-reliant and achieve a good standard of living. First it was Britain. The invention of the steam engine and the Industrial Revolution turned rural Britain into Great Britain. Likewise from the late nineteenth century onwards, manufacturing, mass production and scientific management changed the United States from being an agricultural nation to being an industrial world power. And it is manufacturing, and the pursuit of quality in manufacturing, that has made Japan the strong economic nation that it is today.

Thus the wisdom of the shift of focus from service back to manufacturing as the competitive weapon in the global market cannot be faulted. The concern now is that the West, confused by the huge success of Japanese manufacturing industries, is almost blindly trying to follow Japanese manufacturing practices. High profile consultants market Japanese practices in the form of a host of three-letter acronyms (TLAs), such as TQM, TPM, TCI, JIT, etc. The consultants are careful to point out that the TLAs are not a quick fix. But by the very act of warning that TLAs are not 'quick fixes' they subtly imply that, if their firm is consulted, a quick fix can happen. Of course there are some success stories, but there are many failures and a growing cynicism amongst management and workers alike.

Another concern is that getting the factory right (manufacturing excellence) without addressing the whole spectrum of the business, internally and externally, is of little benefit. The achievement of manufacturing excellence just at the factory level will only result in a fraction of the competitive advantage that could be gained.

The argument is that there should be a 'marriage of marketing, manufacturing and marketplace' (Shapiro, 1988). Likewise the 1990 *European Manufacturing Futures Survey* by De Meyer and Ferdows (1990) showed successful European manufacturers are 'focusing increasingly more on establishing closer links between production and other functions in the company as well as suppliers, customers, and others outside the company'. Wild (1995) sees the relationships between manufacturing, marketing, personnel, finance and design as 'basic factors', and Schroeder (1993) says, 'In the past there has probably been too much attention in operations to internal efficiency. Operations has been asked to concentrate on internal improvements and to leave the external concerns to others.'

A detailed study of 14 blue chip UK based companies found that there are 17 success factors to being a world class manufacturer, and that these factors cover the whole spectrum of a business (ICPG, 1989).

The third area of concern is the reliance on 'benchmarking'. That is the process of measuring a company's own business practices against competitors or industry leaders. There have been many benchmarking exercises to establish the position of a company against the 'best practice'. However our experience is that unless members at all levels of a participating company believe the business can benefit, the exercise has little value. If a company believes that they already know the best way, or that the 'best practice' is not appropriate to their circumstances then improvement will not happen.

If adopting Japanese practices in the form of a series of TLAs, relying on the factory alone to achieve manufacturing excellence, and benchmarking to see how we stack up will not get us there, what will? Having posed the question we then debated the real issue, which is, 'How can a manufacturing business achieve the competitive edge in the global market?' We concluded that a simple systematic procedure for evaluating all aspects, company wide, of manufacturing standards in order to identify areas of improvement was needed. To meet this need, total manufacturing solutions was developed. Total manufacturing solutions produces a total 'score board' of manufacturing and identifies the gaps. Once the gaps are known then corrective action can be taken.

What is total manufacturing solutions?

A conventional definition of manufacturing is a process of converting materials, energy and information to a product or services on a large scale, for customers. The conversion process takes place in a 'factory' comprising facilities (hardware), procedures (software) and human resources.

We have defined total manufacturing in a wider sense. Total manufacturing is concerned with the interactions between the conversion processes inside a 'factory' with all other business processes such as marketing, research and development, supply-chain management, financial and information management and human resources management – and also with the external factors such as environment and safety and customer care and competition.

Total manufacturing solutions is the systematic process of measuring all aspects of manufacturing business against 20 defined areas ('foundation stones'), and of identifying areas of improvement for achieving the full potential of the business. Answers are sought to specific questions and then, depending on the level of performance, scores are allocated. The aim is to find the correct balance of 'foundation stones' for building a particular manufacturing business. The business is built from the foundation stones up and consists of six 'pillars' of total manufacturing solutions. The pillars are:

- marketing and innovation,
- supply-chain management,
- environment and safety,
- manufacturing facilities,
- procedures, and
- people.

The process comprises five steps. The steps are:

1. Define and understand the 'pillars' and the corresponding 'foundation stones' of total manufacturing solutions.

2. Establish a questionnaire for each 'foundation stone'.
3. Obtain data, measure the manufacturing correctness factor and map the performance level.
4. Analyse gaps, both against the best practice (or world class) and the business mission.
5. Implement measures to minimize the gap and thus improve competitiveness.

The self-assessment approach of total manufacturing solutions provides a sound basis for internal comparisons and allows a company to measure and improve progressively to its full potential in the future. It will also enable a company to consider independent external assessments and seek expertise in weak but priority areas.

Pillars of manufacturing solutions

The success of a manufacturing business rests on the six pillars of manufacturing solutions. Each pillar relies on specific and defined 'foundation stones' as shown in Table 1.1 and as illustrated in Figure 1.1.

Table 1.1

Pillars	Foundation stones
1. Marketing and innovation	1. Understanding the marketplace. 2. Understanding the competition. 3. Product and process innovation.
2. Supply chain management	4. Manufacturing planning and working with suppliers. 5. Distribution management and working with customers. 6. Supply chain performance.
3. Environment and safety	7. Product safety. 8. Industrial safety. 9. Environmental protection.
4. Manufacturing facilities	10. Sourcing strategy. 11. Appropriate technology. 12. Flexible manufacturing. 13. Reliable manufacturing. 14. Manufacturing performance.
5. Procedures	15. Quality management

		16. Financial management
		17. Information technology and systems.
	6. People	18. Management skills and culture.
		19. Flexible working practices.
		20. Continuous learning.

Figure 1.1 shows the 'structure' of manufacturing solutions.

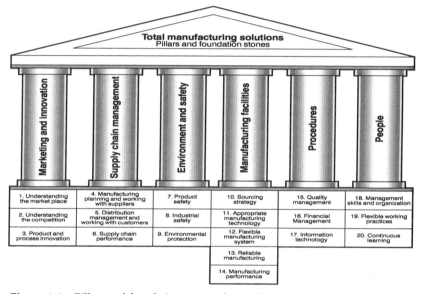

Figure 1.1 *Pillars and foundation stones of manufacturing solutions*

Self-assessment by 200 questions

Self-assessment is by 10 questions for each of the 20 foundation stones (200 questions in total). The questions are easy to interpret and depending on the answer, a score can be allocated as follows:

0.1 Poor   0.4 Very good
0.2 Fair   0.5 Excellent
0.3 Good

For some questions there should be a quantifiable range of measures (e.g. operational efficiency of a production line: below 40 per cent = poor, over 80 per cent = excellent).

Thus the maximum score for the 10 questions of each foundation stone will be five and the total possible maximum score for the 20 foundation stones will be 100.

Data recording

After the completion of the assessment the company will be able to record scores for each 'foundation stone' (and in doing so for each 'pillar') and thus a total score will be obtained.

The total score (out of 100 maximum) is the manufacturing correctness factor. A 'spider diagram' can be constructed from the scores of each foundation stone to highlight the manufacturing performance profile. An example of a spider diagram is shown Figure 1.2.

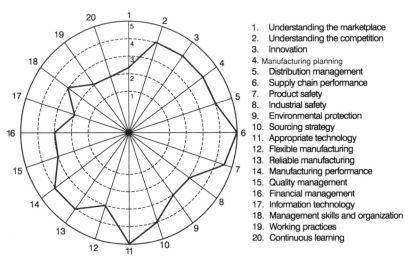

Figure 1.2 *Manufacturing correctness profile ('spider diagram')*

Gap analysis and mission statements

Once data analysis has taken place and the manufacturing correctness factor determined for each pillar, based on foundation stones for the pillar, then gap analysis can take place. However, before gap analysis is attempted it is important that the company's mission is clearly stated and understood.

The fashion these days is for every company to have a 'mission' and every new chief executive appears to feel obliged to issue a new mission statement. (There is some evidence that a new chief executive will see the issuing of a new mission statement as a method of establishing his or her authority.) It is important that we understand the true mission of a company. The true mission may in fact have little in common with what the published statement says. The statement may be mere rhetoric, or words 'full of sound and fury, signifying nothing'. For example if the true mission (in tight economic times) is survival, then it matters not at all if the mission reads, 'to be world leaders by being customer focused'. Of course one could not expect a company to publish a statement that reads 'Our mission is survival', but if this is the case then it is essential that the true mission is understood, and defined, before gap analysis is attempted.

Having established the true mission, it is then equally important to establish the strategic goals of the company. In this sense the strategic goals are those steps that have to be taken to make the mission happen. The mission tells us where we hope to be going, and the strategic goals tell us how we will get there. Once the mission and goals are established, then the relative importance of each pillar to the business can be assessed. It could be that all the pillars are of equal importance!

The gap analysis begins with the manufacturing correctness factor, then moves to the important pillars to the business (as defined against the mission) and their foundation stones. For example a low manufacturing correctness factor such as below 50 would signify an overall weakness in the manufacturing business. If product innovation was essential to the mission then the marketing pillar and foundation stones 1, 2 and 3 (as shown in Figure 1.1) would become the focus of attention.

Improvement process

The improvement process will depend on the magnitude of change required. Where a major shortfall is identified then the focus will be on the immediate need to fix that area. This is known as focused improvement. Where less drastic changes are required the aim will be for continuous improvement.

Measures for focused improvement will apply when changes in cost, time or quality must be instituted rapidly. Continuous improvement measures are progressive and measured, and the process of improvement is over a long-term programme.

Summary

In this chapter it was established that only through manufacturing will a company gain a competitive edge in the global market. It was determined that there is a need for a simple and all-embracing method of achieving manufacturing excellence. The method introduced in this book is total manufacturing solutions. The total manufacturing solution method uses 200 structured questions which will enable self assessment by a company of its strengths and weaknesses. The analysis of the questions is at three levels: manufacturing correctness factor, pillars and foundation stones. Particular emphasis is placed on identifying the 'true' mission. Gap analysis is concentrated on the pillars and foundation stones which are most important to the achievement of the mission.

References

De Meyer, A. and Ferdows, K. (1990) *Removing The Barriers In Manufacturing.* INSEAD.
ICPG (1989) *Manufacturing 2000.* Inter-Company Productivity Group, John Russell Associates.
Schroeder, R.G. (1993) *Operations Management: Decision Making in the Operations Function.* McGraw-Hill.
Shapiro, B.P. (1988) What the Hell is 'Market Oriented'? *Harvard Business Review*, Nov-Dec, 66, No. 6.
Wild, R. (1995) *Production and Operations Management.* Cassell.

2

Understanding total manufacturing solutions

> *O wad some Pow'r the giftie gie us*
> *To see oursels as others see us!*
> Robbie Burns

Our theme is that manufacturing creates wealth through the creation of a tangible product. In Chapter 1 we outlined our method of self-assessment through 200 questions to achieve competitive advantage in manufacturing.

The aim of manufacturing is the creation of a product. Creation of a product occurs through the transformation of raw materials into a finished article. The transformation process uses resources of people, materials, and capital (in the form of plant, machinery and buildings). The efficiency of the transformation process is dependent on the accurate and timely flow of information. So much is self-evident. Our concerns are two-fold; first, with the efficient use of resources and the elimination of activities that do not add value to the process, and second, with understanding and streamlining of external and internal flows of information. Externally between factory, supplier and customer, and internally between the various functions or departments of the organization.

Why 200 questions?

The six pillars and their foundation stones, and our 200 questions are designed to show the way to efficiency in the total supply chain, the elimination of unnecessary expense, and the simplification and improvement of the flow of information.

We pose the questions. The questions diagnose the problems and opportunities. The answers are dependent on individual circumstances. We believe that if the right questions are asked then the answers will often become self-evident. Knowing which questions to ask is part of the answer, but it is also important not to be fooled and to accept the first glib easy answer that we get.

The Japanese have an approach known as the five whys. Davidow and Malone (1992) give an example of this. Suppose a machine stopped functioning:

1. **Why** did the machine stop?
   There was an overload and the fuse blew.
2. **Why** was there an overload?
   The bearing was not sufficiently lubricated.
3. **Why** was it not lubricated sufficiently?
   The lubrication pump was not pumping sufficiently.
4. **Why** was it not pumping sufficiently?
   There was no strainer attached and metal scrap got in.
5. **Why** was no strainer attached?
   There was no preventative maintenance schedule.

Repeating **why** five times in this manner will help determine the root problem so that corrective action can be taken. If in this example the five whys had not been asked the fuse and/or perhaps the pump shaft might have been replaced. If these were the only actions taken, then the problem would recur in a few months. The objective is to eliminate the root cause rather than to patch up the effects (see also Imai, 1986). The five whys is a variation on the classic work-study problem-solving approach of why, what, where, when, who and how. For a discussion on this approach see Wild (1995).

Although we provide 200 questions, each of these questions is not in itself final. Each question should suggest or indeed trigger further questions. We also appreciate that not every organization will find that the six pillars and their foundation stones, and all of the 200 associated questions, will be totally relevant to their situation. Nor do we suggest that each organization should slavishly attempt to apply each of the foundation stones to their own company.

Which type of manufacturing?

Our approach is relevant to any type of manufacturing organization. In any event we believe that the categorization of manufacturing into types of processes is perhaps no longer relevant.

In the past, textbooks have categorized manufacturing processes into job, batch or assembly line. However distinctions have become blurred. For example with the just-in-time approach, supposing a traditional assembly line is still in use, but whereas previously the products were once made in batches of 100 (so as to give what was known as benefits of scale), now on the same assembly line work might be scheduled in batches of one. In one organization we visited, which makes goods products (refrigerators, cookers, washing machines, etc.), there are 3000 different line items and 900 units are made a day. Previously production was scheduled in batches of no less than 100 units; today batches are scheduled in units of one. Previously it took up to eight hours to change the line over and to set up for a batch; today the change-over and set-up time is down to less than three minutes. Another example is with Toyota where they are working towards the concept of a 72 hour car. The idea is that the purchaser will visit a showroom and be able to see a car indicative of the type of product that Toyota makes. There will not be a wide range of vehicles to inspect; instead the purchaser will be shown on a computer screen the various models available and a list of optional specifications. The purchaser will then select, by keying into the computer, the basic car model and required details such as size of engine, type of transmission, colour scheme, type of upholstery, sound system and so on, but all chosen from a given list. This information will now be electronically transmitted to the factory and to the suppliers of the factory. Within 72 hours the car will be delivered to the purchaser. That is the concept. The benefits include the customer getting what they want. But in fact the customer is now more than just a customer; the customer is now very much part of the manufacturing process. In effect by keying in their requirements the customer initiates the whole process, raises the raw materials order for the factory, and updates the production schedule. From Toyota's point of view there is likely to be a further substantial benefit. Presumably the purchaser will be expected to pay on delivery, so there will be no cash-flow problems (within a 72 hour period it is unlikely that Toyota will have paid for the materials or for the direct labour).

Does it matter then how we would classify a manufacturing process? It could be said that in Toyota's case they are using a production line to make one-off jobs.

The relative importance of each pillar is likely to be different depending upon the types of manufacturing processes and products. However, regardless of whether the business is concerned with manufacturing commodity products or consumer products, or whether it is a continuous chemical process or a discrete metal cutting operation, each business has a supplier and a customer. Thus the supply chain exists for all businesses. By applying the same rationale, each manufacturing company, at least to some extent, is involved in activities represented by our six pillars. For example our first pillar, marketing and innovation, has the foundation stones of:

- Competition,
- The market place, and
- The innovation of new products and processes.,

What manufacturer can totally ignore any of these issues? Likewise with the other five pillars:

- Supply-chain management,
- Environment and safety,
- Manufacturing facilities,
- Procedures, and
- People.

We have no doubt that our six pillars, and each of their foundation stones, are of enduring relevance to all manufacturing businesses.

Generally in this book we relate to fast-moving consumer goods (FMCG), on the basis that the concepts and problems associated with this type of production are readily applied to all types of manufacture.

Fast-moving consumable goods are those goods that are consumed, that is generally they can only be used a few times, and in some instances only once. For example we would consider a BIC ball point to be a consumable, but a £100 Parker pen could be said to be a durable. But the overall approach to the manufacture of either might well be very similar. In other words the lessons learnt in the production of a fast-moving consumable BIC pen might be equally applicable to a Parker. However if we compare a BIC throw-away razor with an electric razor, although they are designed to do the same job, one is definitely a consumable and the other is a durable.

Notwithstanding the differences in consumables and durables, we would argue that by concentrating on consumables, and perhaps considering your product, prestigious though it might be, as a consumable, you will get a more direct focus on the complete supply chain, from supplier through to the end user or consumer.

Total business process

The pillars and foundation stones are designed to give us a framework for thinking about our organization and the total business processes. We have not written a textbook on each pillar and we would certainly not claim that a slavish following of the 200 questions is a blueprint for success. Each organization is different and each organization has different problems and different priorities.

Each pillar can be considered alone. It will be noted in Figure 1.1 that each pillar is shown as standing on its own because each pillar can be worked on independently. They are inter-related but are not rigidly interconnected. In this sense we have not devised a systems dynamic approach (see Forrester, 1961). If one pillar is disturbed it may not necessarily affect the others, although if one pillar is weak then the whole structure will be vulnerable. Our approach enables the identification of weak pillars. Action can then be taken to correct weaknesses and to reinforce.

None the less, it is important that a 'big picture approach' is applied and all the pillars of the business are examined. The synergy that results from the benefits contributed by all elements as a whole, far exceeds the aggregate of benefits given by individual elements. The integrated approach is truly more than the sum of its elements. If one concentrates exclusively on isolated areas, a false impression may be inevitable and inappropriate action taken.

This maxim can be illustrated by the Indian folk tale of four blind men who were confronted with a new phenomenon, an elephant! The first man, by touching its ear, thought it, the elephant, was a fan. A second was hit by its tail and concluded that it was a whip. The third man bumped into a leg and thought it was a column, while the fourth on holding the trunk decided that it was an over-sized hose. Each man, on the evidence he had, was right in his own way, but all had made an erroneous judgement by failing to deduce that the total object was an elephant. As with all feedback devices where a basic message is given, inferences and decisions may be drawn from isolated data which will be false and misleading.

A fairly recent television advertisement for a quality British newspaper showed a photograph of a young, scruffily dressed man running towards an elderly gentleman. The viewer was asked what was the young man doing. The obvious inference being that his intention was menacing and possibly an attack was imminent. What the camera revealed in this scene proved to be only part of a larger photograph. The whole picture showed that the older man was walking beneath a building site unaware of a large heavy sack falling towards him and the younger man was trying to warn him.

A story in the business context will further underline the limitation of tackling only a part of a total problem. The technical director of a multinational company, having been to a conference, decided that line performance improvement must be the best thing in manufacturing. So he organized his technical team, called in experts from the corporate headquarters, and set up a line efficiency exercise. The team did an excellent job on two production lines by systematically eliminating all machine related downtime problems (with the aid of highspeed video techniques). As a result the production efficiency of the lines increased by 20 per cent. However, it soon transpired that the product for one of the lines was going to be discontinued and the other line, despite its excellent standard of reliability and efficiency, encountered a severe long-term shortage of materials due to planning and procurement problems. Therefore, in isolation the line efficiency programmes did not improve the overall business performance.

Own benchmarking

There are many proponents of a 'single' method of problem solving, usually designated with a buzz word or a three letter acronym, whether it be called business process re-engineering, lean production manufacturing or whatever else the flavour of the month is. Consultants can get away with this kind of one track package as long as their market (that is you and your management), is relatively ignorant of appropriate options. Over the past decade, however, business people have become increasingly aware of the need for company-wide methods and approaches, perhaps encouraged by the success of Japanese manufacturing industries. The art of management has grown much more widespread. One cannot therefore blame managers if they now make cynical remarks when they come across a 'packaged' approach from consultants.

Our recommendation is that managers need to do their own benchmarking. To do this self-assessment they need a comprehensive method of benchmarking, and our intention, with our 200 questions, is to provide such a comprehensive approach. Once a weak point is diagnosed a specialist or consultant may be employed to effect a cure or act as a catalyst.

Our approach is a diagnostic process which depends on 200 questions. Therefore it is important that we delve into the background of these questions. In Chapters 3 through to 8 we examine each of the six pillars and their foundation stones so that a total manufacturing solution can be constructed, just and perfect in all its parts.

What is required of you is to consider yourself a new breed of manufacturing manager. Your first step in this new guise will be to familiarize yourself with the six pillars and 20 foundation stones (Chapters 3 to 8) in the context of your business. Then you should be able to benchmark your own business by using the 200 questions (Chapter 9). The methodology of data collection and analysis is given in Chapter 10. Once this is done it is likely that changes will be necessary, which could well entail a change in culture. Chapter 11 deals with how a well constructed mission statement can help effect a change in culture.

We have found that many mission statements are poorly worded.

Properly used, the mission statement should define exactly what is required. Chapter 11 includes a section on how to create a meaningful mission statement. Words such as to be the best, to be world leaders, are not words we favour. Rather we see the mission as being a statement of exactly what we want to do. A well worded mission gives a focus for strategy.

For example, supposing the vision was to set up an ambulance service in a small town. The mission could well read:

> 'To provide a speedy response transport and first aid service for the sick and injured.'

Note the brevity of this statement, no mention of to be the best, or to be world leaders or to be professional, etc. All these attributes might be highly desirable but for a succinct mission statement, getting back to basics will help crystallize exactly what we are trying to achieve.

From this statement we can then list what will be required to make the mission happen. In this example:

- **To provide a speedy response.** This would mean we would need a 24 hour service, and we would thus need to have staff on call 24 hours a day. A communication system will be necessary. Ideally, to give a speedy response, a central location would be desirable.
- **Transport and first aid service.** This suggests reliable vehicles and equipment, and competent staff. If we are to perform then the vehicles and equipment will need to be well maintained, and the staff will need to be well trained. Maintaining and training will need to be ongoing.
- **For the sick and injured.** This reminds us why the service exists and defines who our customers are.

To make our mission happen we would need to consider the following list of 'things to do':

1. Select and acquire vehicles
2. Maintain vehicles
3. Select and acquire equipment
4. Maintain equipment
5. Recruit skilled staff
6. Ongoing training of staff
7. Select a location
8. 24-hour service, staffing and communication
9. Network with hospitals and other emergency services

16 TOTAL MANUFACTURING SOLUTIONS

The above is meant to be a simplified example. Likewise with our approach in this book. It is not for us to be experts on your organization. We provide a framework for you to apply. From our own practical experience we can testify to the effectiveness of our approach.

Your own benchmarking, using our 20 foundation stones and 200 questions, will position your business against a backdrop of criteria comprising both performance and practices. The exercise cannot have any practical value to your business unless the general profile can be matched with the specific mission and objective of your organization. In Chapter 12 we provide a methodology which allows the scoring of each foundation stone to be 'weighted' according to individual business priorities. The manufacturing correctness profile will identify the gaps in your business and show where further improvements are needed.

Chapter 13 discusses improvement strategies, and Chapter 14 shows a plan of how to start the improvement process and how to make things happen. In Chapter 15 we discuss the totality of our approach. We conclude that total manufacturing solutions is for the whole company and cannot be limited just to the factories.

Summary

To summarize we take a practical application based approach. However we are not simply propounding 'back to basics'. We are providing a conceptual framework to enable organizations to rethink what they are doing and why. The aims are two-fold: first, the elimination of non-value-added activities, and second, the improvement of the flow of information throughout the total organization (internally), and externally with and between, suppliers and customers. Our overall approach is to examine each pillar and each foundation stone for weaknesses but not to ignore the whole by concentration on just one area. Our intention is to allow you construct a manufacturing organization which is just and perfect in all its parts.

References

Davidow, W. and Malone, M. (1992) *The Virtual Corporation*. HarperCollins.
Forrester, J. W. (1961) *Industrial Dynamics*. MIT Press, reprinted 1969.
Imai, M. (1986) *Kaizen: The Key to Japan's Competitive Success*. McGraw-Hill.
Wild, R. (1995) *Production and Operations Management*. Cassell.

3

Marketing and innovation

Two roads diverged in a wood, and I -
I took the one less travelled by,
And that has made all the difference.
                                      Robert Frost

It is said that today the priority for top management is to deliver a quality product to meet and even to exceed customers' expectations. This has to be achieved within the constraints of what the organization can afford to provide and is capable of providing. Shareholders and investors still measure management's performance in terms of the bottom line. On the other hand, customers, driven by intense international competition coupled with the promises of technological advances, are continually setting higher levels of quality and service, all at less cost. This chapter deals with maintaining a competitive advantage through innovation and marketing.

Generally businesses are organized around functions. Even in organizations that have re-engineered, and no matter how flat the structure, some people's prime function will be marketing and some people's prime function will be manufacturing. Others will be involved in accounting, administration, human resources and so on. Our main concern in this chapter is with the marketing and manufacturing functions.

The essential job of the marketing function is to define a product that will sell. Definition of the product includes the features or attributes the customer wants from the product. Attributes may range from the absolutely essential, through to the desirable, down to the lower level of nice to have but not really important. Attributes can include the finish and aesthetic appearance (to some customers appearance and status of a product can be every bit as important as the performance). Often some of the features required by the customer arise as the result of marketing pushing (selling) features that the customer had not previously considered important. As well as defining the product, marketing also has to establish the price the customer will be prepared to pay, plus the likely level of demand. The customers' expectations of attributes, quality and price, will be driven by their perception of what they have previously experienced, what they believe the competition can now offer, and what they expect will be available (technological advances) in the near future. Often what the producer sees as being state of the art can well be seen by the customer (beguiled by technological promises and exaggerated promises of competitors), as simply old hat.

Once marketing believe that they know what is wanted and what will sell, then manufacturing has the problem of determining feasibility. That is to say, can the product with all the desired features, actually be made? Does the business have the technology (the know how and the specialized plant or equipment), and does it have the capacity (what else has to be produced and what are the priorities)? If, after taking all the above into account, and it is decided that it can be done (there is sufficient capability and spare capacity, and thus the quantities required can be made within the time frame advised by marketing), then the crucial question must be 'is it possible to make to the price set by marketing?'

Thus arise the traditional conflicts of marketing and manufacturing. Marketing see themselves as the opportunists, they are the innovators, the go-getters, they are the trigger for making things happen. For them the bottle is always half full whereas they believe that their colleagues in manufacturing would describe the same bottle as half empty! Manufacturing see themselves as the realists, they also see their job as making things happen, but it is their responsibility to balance conflicting demands with scarce resources. They know only too well who will get the blame when an order is delivered late and not to specification!

Thus marketing often see the response of manufacturing as being negative. They believe manufacturing looks for reasons as to why things can't happen rather than looking for ways in which to make things happen. On the other hand manufacturing see that marketing has little or no appreciation of the problems of capacity and scheduling, and the time and effort required to develop new products. Then again marketing always seem keen to add to the range of product lines or stock keeping units.

These conflicts are highlighted in a question/response type scenario as shown in Table 3.1.

Table 3.1

Marketing	Manufacturing
Why do we never seem to have enough capacity?	We need accurate long-range forecasts.
Why are our lead times so long?	We need accurate medium-term forecasts.
Why do we have stock outs?	Why is your short-term forecasting so erratic?
Why are our costs so high?	Extended product lines, rushed deliveries, fancy extras, and high quality finish all cost money.

... and so on, until the exchange escalates to the extent that it can be likened to children throwing stones at each other across a garden wall (with each side determined to throw a bigger stone or brick back each time). We are sure that from the reader's experience other similar questions and replies will spring to mind. The pointlessness of such an exchange is obvious.

Part of the problem is communication and lack of understanding by both functions. Much of this is due to the traditional hierarchical structure of organizations whereby functions are 'walled' off from each other. This can be described as the bunker mentality whereby each function sees the other as a possible threat or a challenge. In this atmosphere of mistrust responsibilities and demarcations are jealously guarded and suggestions, however helpfully meant, can be seen as examples of meddling or trying to usurp authority.

If we want to do it right we need total manufacturing solutions. This requires not only bringing all people together 'by breaking down the walls' but also a good understanding of:

- the marketplace,
- the product innovation process, and
- the competition (who they are and what they are doing).

If we return to Chapter 1 we will see that the pillar for marketing and innovation comprises three 'foundation stones':

1. Understand the market place
2. Understanding the competition
3. Product and process innovation

We will now discuss each of these foundation stones.

Foundation 1: Understanding the marketplace

Some managers will tell you that they thrive on chaos and that they enjoy the challenge of crisis management. There is a place for people like that: ideally with the competition. Certainly it is important to be able to quickly react to threats and problems as they arise, but in reactive type situations the best possible result is seldom achieved. Surely it is far more desirable to be in a position to anticipate what is going to happen and to plan accordingly? The ideal situation is to be the market leader, and for you to be making things happen, and to put your competitors in the position of playing to catch up.

The first stage of developing any business, strategy, or improvement plan is to analyse and understand the nature and trend of the market you are in. To be in a position of understanding what is happening and what is going to happen it is necessary to know:

- the size and trend of the market,
- customer perceptions,
- distribution channels, and
- global opportunities.

In order to obtain a better knowledge of the size and trend of the market, the starting point can be trade, consumer and government statistical publications. It is also important to understand the local demography, the culture and habits. At a basic level if you want to open a fast food business you do not use pork in Israel or beef in India. At a more subtle level, if you want to market a brand image (e.g. Gucci) a large population alone (e.g. China) does not determine the market size. A growing market offers opportunities for future profits if a dominant position can be obtained.

For a given cost of a product, a customer's requirements will fall into three categories: time, quality and service. Time includes the lead time for new products, i.e. the time between the start of the product definition and the time when the first shipment is made. For existing products time means the lapsed time from the date of making the order to the date of delivery. Quality includes the defect level of products as perceived by the customers and quality also includes delivery on time. Service is as judged by the customer.

It has been said 'if it can't be measured it can't be managed.' Some measurements are internal and some external. Internally, on the shop floor it should be a straightforward task to record and measure the cost of scrap, reworks, downtime of machines, and idle time of workers. Likewise delivery dates should be known and it should be standard practice to record the percentage of deliveries made on time. Measurement of a service is not so easily achieved from within, but the percentage of deliveries made in time to customer specification is one measurement that can be made.

There are some other measures and ways of determining how your customer perceives your service levels. First establish who your top 20 per cent of customers were last year. (Statistically about 80 per cent of your total sales will be from 20 per cent of your customers.) If marketing doesn't have this information, the accounts department will easily be able to extract it. Now compare this year's sales against last year's. Have sales dropped for any of the major customers? If so go and talk to them. If approached in the right manner (a genuine desire to improve quality and service, and an avoidance of excuses) then customers usually will only be too pleased to tell you what you are doing wrong. If there is some doubt that you can perform, be careful of making promises to 'get it right' the next time. Failure to get it right next time will totally destroy the credibility of your organization!

It is also important to recognize who your real customers are. Are they the agents, the wholesalers, the retailers or the end consumers? Analysis of the distribution channel will help to determine exactly who the customer is and who is driving demand. Once this is known, then by talking to the real customer, it can be found what attributes are really required of the product. How important is the finish, the price, and so on? By following the distribution channel back to the end user, one manufacturer found that a costly feature was never used by the end consumer and thus could be removed from future models. (It had been added to meet a whim, albeit advised in an authoritative manner, of the wholesaler.)

It is also important to be aware of potential customers. For example the emerging markets of what was 'the communist block' and the gradual removal of tariff barriers have all resulted in a huge growth in world trade and for most companies this has opened up opportunities in the global marketplace. Conversely an opportunity can just as easily be a threat. As Creech (1994) writes, 'Go to the annual Consumers Electronics Association Convention in the United States ... Japanese, Koreans and others from Oriental countries so overwhelmingly dominate the sales personnel and display booths it's as if some exotic malady has wiped out the American Suppliers'.

Foundation 2: Understanding the competition

Having established a good understanding of the nature and trend of the marketplace, the next stage is understanding your market position and the position and strength of the competition in the market. Activities include:

- establishing your 'core' business,
- analysing your product portfolios,
- identifying the competitors, the setting of performance criteria based on the competition, and the measuring of your current performance against these criteria, and
- 'SWOT' analysis (strengths, weaknesses, opportunities and threats) to compare your business, internally and externally, against the competition (current and possible competition).

Core business

The first step in establishing the 'core' business is to examine the product range and roughly establish the margin or profitability of each product or group of products, and then look for products within this group that have 'differentiation', i.e. the products which are perceived to be superior to those your competitors are marketing. Differentiation can take many forms, such as:

- brand image, e.g. Coca-Cola
- technology, e.g. Sony Camcorder
- service, e.g. Singapore Airlines

The 'core' business may also evolve as a subset of categories as a 'focused' strategy to a restricted area of the market. An example of this strategy is Porsche with a niche market in prestige sports cars.

Product portfolios

The analysis of establishing the 'core' business into categories is, however, extremely general. The next stage is to convert these generalities into specifics by further analysis of product portfolios. (A product portfolio is a grouping of products which compete in the market place in identical ways e.g. a London–New York air ticket.)

There are several measures for analysing the relative importance of a product portfolio including contribution, market share, growth of market share, and market growth. But seldom can one measure be taken in isolation (a product with an excellent margin is not core business if only a few items are sold each year). The importance of contribution, market share, growth of market share and market growth is discussed below.

- **Contribution:** Contribution or profit margin of a product is an indication of how potentially profitable a particular product portfolio is.
- **Market share:** If your product has a good profit margin and a high market share, especially if the product is a price leader in the market, then you are in a very strong position.
- **Growth of market share:** If the product's market share is increasing, particularly in a growing market, and it is the price leader, then the product has a very real competitive advantage.
- **Market growth:** This is an indication of future opportunities for the product portfolio.

The analysis of product portfolios is more complex than the simplified description of the above measures.

One popular and established approach is the BCG (Boston Consulting Group) matrix of 'stars', 'wild cats', 'cash cows' and 'dogs' against the

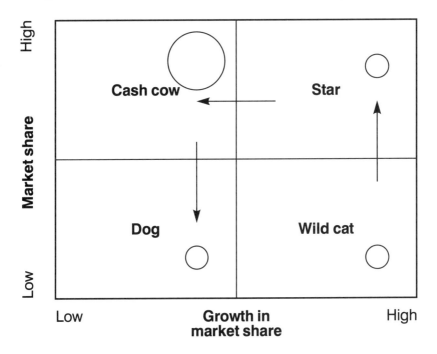

Figure 3.1

two axes of market share and growth as illustrated in Figure 3.1:

- A **cash cow** is a product with high sales but which is static in market growth. With a cash cow, because good sales are still being made, the danger is that unless market share is known and is being monitored it will not be realized what is happening. With a cash cow unless action is taken, for example the addition of new features or a new model, then the product is likely to become a dog.
- A **dog** is a product with a low share of the market and with no growth in the market share.
- A **wild cat** can become a star. A wild cat has a low share of the market but its share is growing.
- A **star** has a high share of the market and its share is growing.

Although the BCG matrix is a useful way to obtain a quick picture of the position of the product portfolios, it is often difficult to collect reliable data of the market share. Another approach is to carry out a similar analysis of products into A, B, C or D against two axes, one of contribution and one of growth, as illustrated in Figure 3.2.

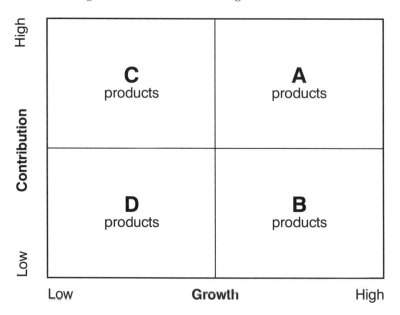

Figure 3.2

A products have a high contribution and their sales are growing. At the other end of the scale D products have a low contribution and their

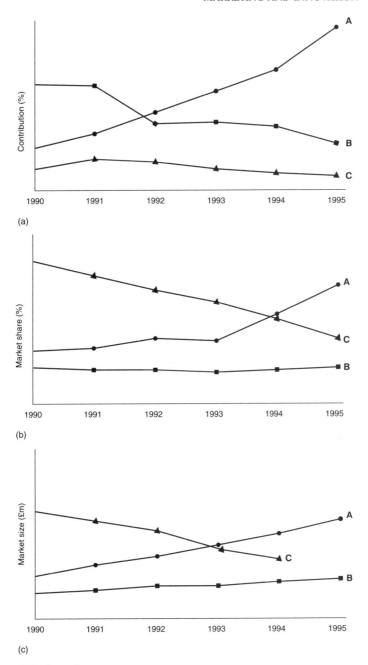

Figure 3.3a, b and c

sales are not growing. For B products, although contribution is low, growth is high, and with C products although they have a high contribution their market growth is low. For C products the question is: can the margin be sustained in a falling share of the market? Without a change in product features the only way to increase market share will be to reduce the price, but if this doesn't work then not only has contribution been reduced but the product is likely to slide down into the D category.

Information for the above analysis is easier to obtain and is more accurate than the BCG analysis because all the information required is available within the company. However this does not provide a direct measure of competition.

Another useful approach is to plot trends in per cent contribution, per cent market share, and market size (£, tonnage or units) over a number of years, as illustrated by Figure 3.3, Figure 3.4, and Figure 3.5, respectively. Once this information is plotted and analysed, then the core business of important products should be readily identifiable.

Having identified the important product portfolios, the next stage is to identify competitors in each portfolio and the competitive performance criteria.

Competitors

Identification of existing competitors should not be difficult. It is a poor sort of marketing department that can't identify the existing competitors!

Performance criteria will depend on what product and what market you are in. For example for a manufacturer of fast-moving consumer goods (FMCGs) the chosen criteria could include measurements of:

- innovation and success ratio,
- cost, quality and service of products,
- capacity and performance of manufacturing units, and
- efficiency of distribution.

The method of assessing your own performance against your competitors is popularly known as 'external benchmarking'. Establishing a benchmarking partnership with direct competitors may not be easy as it is an unaccustomed situation to 'dance with the enemy'. However the problems of access may not be so acute when the partners to benchmarking are operating in a different market. In some respect businesses in competition have a clear interest in benchmarking each other in common peripheral areas – such as purchasing of commodity type raw materials – and thus are willing to co-operate. Where the competition is not interested in co-operating your own market intelligence and the use of external consultation give sufficient information to establish key benchmark measurements.

Much comparison work can be done without the co-operation of competitors or without resorting to consultants to get the information. Manufacturers of cars have for many years obtained competitors' models and stripped them down to see what the differences are and to determine what advances have been made. In other cases the vehicle may not be stripped down but checked against in-house standards as to how the competition for the same class of vehicle measures up in terms of paint quality, fitting of panels, and so on – the aim being to determine how good 'we' are compared to 'them'. Annual reports, brochures and advertising also provide useful information on the standards that the competitors are publicly setting for themselves. Even if the competitors are not achieving these standards, these are, none the less, the standards that are being advised to the market and consequently shape and form the perceptions by which your customers and potential customers will judge the performance of your business and your products.

SWOT analysis

Armed with the best market information available the next self-analysis that can be carried out is SWOT analysis.

Traditionally SWOT analysis is done from two perspectives, internal and external. Internally we identify our strengths and weaknesses and externally we look for the opportunities in the marketplace, and conversely what are the likely external threats. When we look at our strengths we consider what we are doing well with what we have and what else we could be doing; in other words what advantages do we have. Advantages could include our specialised equipment, our committed work force, and a strong financial backing. Likewise the same approach is taken to identify internal weaknesses. Weaknesses could include a lack of co-operation between marketing and manufacturing. Opportunities are external and as discussed earlier could be the lowering of tariff barriers and the opening up of new markets (of course lowering of tariff barriers could also be a threat). Threats, once again from without, will not only be the competition or likely new competitors but will also include possible legislation, technological advances (which could also be an opportunity) and other issues such as the environment and the green movement.

Foundation 3: Product and process innovation

Having identified the competition, the core business, and the market position for the 'core' business it is now time to assess the 'gap' in the existing product portfolio and to move to improve existing products and processes, and/or to innovate new products with expanded capabilities to fill the gap.

Innovation is essential to keep pace with what the competition is doing. Innovation includes the search for new products as well as the improvement of existing products. Since the number of entirely new products will normally be few, development will largely be the introduction of adaptations, improvements and the addition of features. In this regard the technique of 'value analysis' has proved important. Value analysis looks at existing products, in an organized way, element by element, with the aim of reducing cost without reducing the performance or reliability. For example the replacement of a brass casing with an alloy or even a plastic material might reduce the cost and actually improve the product (now lighter to carry around yet performance has not been reduced and it is much cheaper and easier to produce).

For new innovations Ray Wild (1995), has identified six stages of development. These are:

1. exploration, including research, i.e. the continual search for new ideas;
2. systematic, rapid screening to eliminate less promising ideas;
3. business analysis, including market research and cost analysis;
4. development of the remaining possibilities;
5. testing the offerings developed;
6. launching on a commercial scale.

There are several strategies available to a business with the development of new products. Some, a few, will position themselves as the market leaders and will, through high investment in research and development, bring forth new products. Others will seek quickly to copy the innovations of others and will attempt to join in the initial growth phase of a new product. Others will join in with adaptations before market saturation sets in and will endeavour to perhaps find a niche market. Others will add nothing new to the innovation but will rely on mass production to enter the market at lower prices.

For any organization, whichever of the strategies are adopted for innovation, certain conditions have to be met:

- First, close links between marketing, research and development, and manufacturing are essential.

- Second, innovation lead time has to be minimized.
- Third, there has to be a continuous analysis of the 'product life cycle'.

> Product policy – what to make and how to make it – is the most pressing issue that manufacturing companies face today.
>
> (Shapiro, 1988)

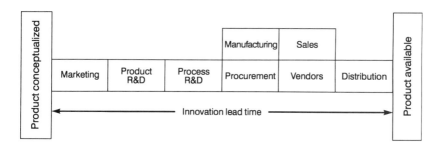

Figure 3.4 *Traditional innovation project*

Business process re-engineering includes the concept of 'integrated product line management' wherein all the major functions – marketing, R&D, engineering, manufacturing, logistics and sales – are involved and the innovation lead time is reduced by interactive and concurrent activities. Figures 3.4 and 3.5 (adapted from examples in *Winning Manufacturing* by J.A. Tompkins, 1989) illustrate the innovation processes by two methods.

It is generally accepted that products have a life which goes through a cycle of launch (incubation), rapid initial growth, followed by a period of maturity and then an eventual and often rapid decline. A new product with new technology, has a 'switch' point when the new takes over from the old technology. An example can be seen with technology associated recorded music when the cassette tape (which had taken over from the 33- and 45-rpm record disks) was in turn superseded by the compact disc. This pattern is shown in Figure 3.6.

30 TOTAL MANUFACTURING SOLUTIONS

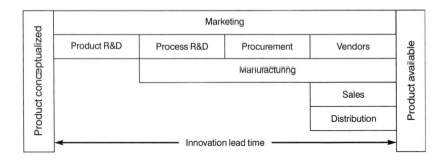

Figure 3.5 *Integrated innovation project*

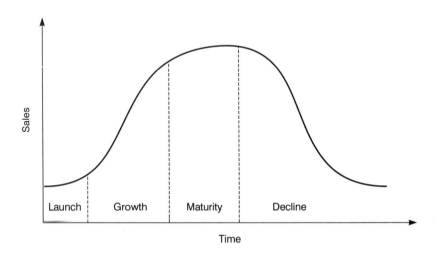

Figure 3.6 *Product life cycle*

Different actions or decisions will be required for each stage of the life cycle, and each decision will have an impact on manufacturing requirements. For example in the launch and the growth stages it will be essential for manufacturing to have the capability to keep pace with the demand, with the maturity stage production requirements will be more predictable and work can be scheduled with some degree of certainty. The decline stage will bring different problems, because to arrest the decline, price changes or additional features might be attempted. Manufacturing will be expected to react quickly and be able to reduce costs or add new features at no extra cost.

Product life cycles for fast-moving consumer goods can be as little as six months whilst for capital-intensive products such as aircraft the cycle can be several years. Often with fast-moving consumer goods even before the launch the next product is in research and development, thus rendering the new product outmoded in a matter of months or even weeks.

Summary

In this chapter we have discussed how customers' expectations are fuelled by what our competition offers or is perceived to offer, and by the promises, as widely publicised in the media, of what technological advances will soon offer.

We say that competitive advantage is hard to maintain, and can only be maintained through innovation and marketing. We agree that the marketing function's role is to know the market place and to be abreast of what the competition is doing. However, marketing is too important to be left to just one section of a company. It is important that all functions, especially manufacturing, be aware of what the competition is doing, and what our key (top 20 per cent) of customers want and expect. Marketing and manufacturing are in it together and must work together, and not throw bricks at each other (bouquets are okay)!

We are firm believers in measurement. Much interesting and important marketing data should already exist in a company, and in this chapter key measures and indicators are considered.

With the subject of product innovation, in this chapter we introduce the concept of 'integrated product line management', and the reduction in lead time by interactive and concurrent activities.

Above all in this chapter, we stress the need for functions to work together and to try and understand and support rather than to criticize or damn with faint praise.

References

Creech, B. (1994) *The Five Pillars of TQM*. Truman Talley.

Shapiro, B.P. (1988) What the Hell is 'Market Oriented'? *Harvard Business Review*, Nov-Dec, 66, No. 6.

Tompkins, J.A. (1989) *Winning Manufacturing*. Institution of Industrial Engineers, Georgia.

Wild, R. (1995) *Production and Operations Management*. Cassell.

4

Supply-chain management

> *The real voyage of discovery*
> *is not seeking new lands*
> *But seeing with new eyes.*
> Proust

To gain a competitive edge, to satisfy customers, and to keep costs down many manufacturing businesses have used the incremental improvement approach of total quality management (TQM) and in more recent times the often drastic restructuring approach of business process re-engineering (BPR). Each of these approaches have their proponents and each approach, or elements of each approach, can result in great advances. Equally each approach has been criticized and there have been many reported examples where an attempt to impose TQM or BPR has resulted in disappointment and even disaster. Some of these disasters can be explained away by saying 'they didn't do it right'. Likewise often when a business is in real difficulties it is too late to hope for a miracle cure. Turning a business around is like turning the *QEII* around – it doesn't happen all at once.

In addition each business is in some way unique. And although there is a tendency for each of us to exaggerate the uniqueness of our business, it is still a fact that what will work for one business will not necessarily work as effectively for another. There can be a variety of reasons why one technique, seemingly successful for one organization, will not work quite as well for another. Reasons can include the type of business, the management style of the chief executive and the overall culture of the organization. Often a change to TQM or to BPR will require a major change in management styles and in the culture of the organization. These issues will be discussed in Chapter 11.

There is however one approach that is becoming increasingly recognized as a certain way of meeting customer specifications and reducing effort and wastage of resources but which doesn't require a major change in direction and culture. This approach is supply-chain management.

Supply-chain management considers demand, supply and inventory needs for each item of production and in particular looks at how inventory flows through the system to achieve output to the customer's specification on time and at least cost. With supply-chain management, customer service is increased through the reduction of lead times and the product is always exactly as specified and it is always delivered on time. Costs are reduced through the elimination of any activity that doesn't add value and through the elimination of any non-essential increment of material.

Activities and measures based on customer requirements, as explained in Chapter 3, are very important in improving business performance. But externally-driven customer-based measures have to be matched by measures of what the company can do (feasibility, capacity, know how and resources) to meet its customers' expectations. A high standard of customer performance derives from planning decisions, processes and actions which take place across the whole organization.

Supply-chain management focuses on the critical measures of all elements of the supply chain. Externally the measures include the suppliers at one end and the customer at the other end of the supply process, and matches these externals with the internal requirements of the manufacturing process. The focus is two-fold; to satisfy customer needs and to keep costs to a minimum.

In reality the elements of supply-chain management are not new – we have all been managing parts of the supply chain for years (e.g. buying, planning, scheduling, stock control, warehousing, logistics, distribution, etc.) without realizing the significance of the whole chain concept. Likewise the cost of the various elements of supply has been long recognized.

> In 50 years between 1870 and 1920 the cost of distributing necessities and luxuries has nearly trebled, while production costs have gone down by one fifth – what we are saving in production we are losing in distribution.
>
> Ralph Barsodi, *The Distribution Age* (1929)

It is, however, new to view the supply chain as a single integrated flow across all the functions of the business. Traditional, specialist functions like purchasing, planning, manufacturing and distribution are substituted by the flows of materials and information across the traditional functional boundaries, as shown in the simplified model depicted in Figure 4.1.

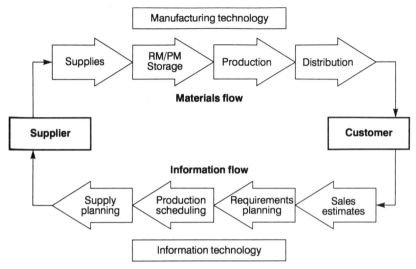

Figure 4.1 *Supply chain management*

Traditionally the information flow was the domain of the commercial division while the conversion process of materials flow was a manufacturing or technical division responsibility. With an integrated supply-chain approach the responsibility for all elements of supply is now with operations management or supply-chain management. In many businesses, the integrated approach is being extended to include all suppliers through the manufacturing processes to each level of customer (including wholesalers and retailers where appropriate through to the end user or consumer). This is known as the extended supply chain.

We will now discuss our philosophy for foundation stones four, five and six.

4. Materials planning, and working with suppliers
5. Distribution management, and working with customers
6. Supply-chain performance

It is on these foundation stones that the pillar of supply-chain management stands.

Foundation 4: Manufacturing planning and working with suppliers

In the past many manufacturers regarded their suppliers with some suspicion, almost as adversaries. Little loyalty was shown to the supplier and consequently the supplier was never certain as to their future relationship with an organization. Often the purchasing, or procurement, department would see their role as screwing the best deal possible from a supplier.

Supply-chain management is distinguished by its role to provide a strategic and integrating function at all levels of logistics including the suppliers. Ideally the supplier becomes part of the team and is involved in the planning process, not only for scheduling of deliveries when required but in the design stage for new products.

In most companies the main objective is to set up a materials requirements plan for inbound logistics so as to achieve an appropriate balance of stock and provide a desired service level to customers.

Materials requirement planning (MRP) is the set of techniques which uses bills of material, inventory on hand and on order data, and the production schedule or plan to calculate quantities and timing of materials. Such a plan is incomplete if it doesn't take into account whether manufacturing resources (e.g. plant, people, energy, space) will be available at the desired time. Manufacturing resource planning II (MRP II) arose from an appreciation of the need to time and phase materials with resource availability so as to achieve a given output date. Manufacturing resource planning II is an integrated computer-based system. A computer-based approach is essential due to the amount of data required. Various software systems are available, each based on the same principles. MRP II is depicted in Figure 4.2.

With manufacturing resource planning the planning process arises from the innovation of new products and the strategic marketing plan. Starting with this information a business plan is constructed to determine and communicate estimates of the sales volume of each product range. The business plan should be developed at least once a year and during the year periodic updates will be required.

From the business plan, an operations plan is formulated which covers the materials and other resources needed to translate the business plan into reality. It follows that to keep the operations plan in line with updates to the business plan, regular communication is required between the various functions involved. This updating process is best achieved by face-to-face meetings which we recommend should take place at least once a month and always with all parties present at the one time. There is a very real danger of misunderstandings and ambiguities if meetings are not face-to-face and if all concerned are not present at the same time. Meetings need not be long drawn out affairs. From experience we believe that any planning meeting that takes longer than an hour is wasting time. The key managers at these meetings will be from sales, operations and planning. The issues that will be agreed will include time and availability of resources, and conflicting requirements

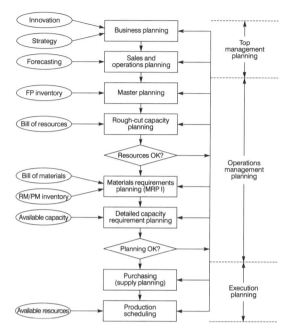

Figure 4.2 *Manufacturing resource planning (MRP II)*

and priorities will be resolved. Above all demand is the crucial issue, and as future demand can never be certain there should be a formal mechanism of forecasting using the best combination of historical models, past results from promotions, data from customers and market intelligence. Likewise the inventory data system has to be up to date and accurate with details of raw materials on hand, goods on order, lead times, and finished goods on hand.

Only with up-to-date information, and with the continuous review and management of information, can an organization hope to achieve a balance of resources and stocks of inventory to meet planned service levels. The master planning and production scheduling process therefore has to be continuously monitored and updated to ensure that this occurs.

The master production plan or master schedule is at the heart of materials requirements planning (MRP) where both the timing and quantity of orders are determined from offsetting from the current stock the demand during the lead time to meet the master production plan.

As shown in Figure 4.3, the concepts of MRP underpinned by the master plan can be extended also to the distribution channel to allow integrated scheduling throughout the supply chain. The approach of distribution requirements planning (DRP) is compatible with MRP as used in the factory.

38 TOTAL MANUFACTURING SOLUTIONS

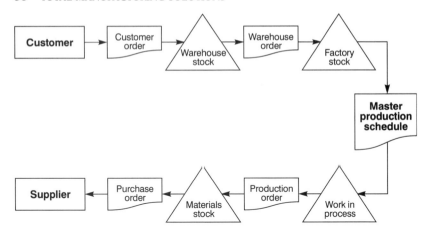

Figure 4.3 *Order flow in materials requirements planning (MRPII)*

The lower line shows actual cumulative sales each week from the start of the planning period. The upper line shows the opening stock plus cumulative production for each week. From the present week (where historical data ends) the current expectation of demand is charted. The minimum stock week offset is drawn from the sales line to derive the minimum production required. The minimum production in any week may then be entered as the cumulative production plan. A similar presentation can be made whereby sales, stock and production can be expressed in units of hours of production.

The next stage is to follow a rough-cut capacity planning process to assess to what extent the capacity of manufacturing facilities could meet the master schedule. The feedback loop at this level tests the master plan against problem areas such as known bottlenecks and other critical resource areas. Often, as this is a short- to medium-term approach, action has to be taken to make the best use of existing resources rather than to add extra long-term resources. The company should decide which alternative to follow if the existing resources are not adequate, e.g. review the schedule, increase resources, work extra shifts, delay maintenance, outsource to third parties and so on. With computer systems it is relatively straightforward to simulate using 'what if' scenarios to evaluate alternative courses of action.

Having established that the resources are sufficient, or having adjusted the plan to fit the resources, then the next step is the detailed materials requirements planning and the detailed capacity requirements planning for day-to-day operations. This stage includes the production of detailed bills of materials for each product or batch of products. With the revised master schedule for each product and for each stock keeping unit (SKU) and bills of material for each SKU, the materials required for each item of raw materials (RM) and packaging materials (PM) are then matched with the current inventory levels to derive the additional procurement requirements. The requirements are modified, if required, after comparing with the detailed capacity planning process.

The execution of the planning process then commences with the final production scheduling and purchasing (supply planning) processes.

To be effective, MRP II has to be an integrated system and should be on-line and accessible to all interested parties. It follows therefore that data has to be kept up to date on the system. For example if engineering changes are made to the design of a product the MRP II database has to be updated otherwise the bill of materials for procurement purposes will not be in line with the new design.

We have outlined a generic description of the manufacturing resource planning process. There are of course variations – more significantly between batch production processes and continuous production processes and between so called 'push' or 'pull' demand systems. With the 'push' system stocks of materials and of finished goods are used to ensure maximum plant capacity utilization by having level production. The 'pull' system is driven by customer orders and just-in-time principles which can result in some under utilization of capacity. It is said that just-in-time requires greater flexibility and reliability of plant plus a multi-skilled workforce. In its simplistic form just-in-time is reactive (demand pull), whereas MRP II can be described as proactive. MRP II looks forward and determines what will be needed to achieve a desired output date. Internally MRP II is a push system; inventory is driven through the process by the schedule. Thus customer requirements are linked to the resources and materials necessary so as to precisely meet a just-in-time delivery date. From a customer's point of view it could be argued that as long as the goods arrive on time and meet the specifications, the system used by the manufacturer is irrelevant!

But a partnership with customers and with suppliers can and will achieve very obvious benefits to all. A partnership with suppliers and a partnership with customers are the beginnings of a radical change in supply-chain management. As a result, the manufacturer, the supplier and the customer achieve benefits in:

- reduced inventory level,
- lower operating cost,
- improved service level, and
- a greater certainty of a continued relationship.

However, the success of the benefits will depend on mutual trust, a highly developed commercial relationship and an efficient system of data exchange. In order to improve the effectiveness of data exchange, companies are sharing with their suppliers (and customers) common systems such as EAN standards (European Article Numbering) and EDI (Electronic Data Interchange). For example EDI enables companies to communicate with each other. Purchase orders to suppliers can be eliminated by using customers' order schedules. And by EDI, the customer could be authorized to link directly into the manufacturer's MRP II system.

Foundation 5: Distribution management and working with customers

In the same way that manufacturing resource planning is concerned with information flow, suppliers and inbound logistics, distribution management is likewise concerned with materials flow, customers and outbound logistics. Inbound logistics is characterized by demand variability, and outbound logistics is characterized by variable service levels. A simple model of distribution management in a manufacturing process is illustrated in Figure 4.4.

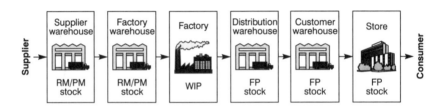

Figure 4.4 *Outbound logistics*

With the management of distribution, that is the physical transportation of goods from the factory to the customer, invariably some stock is held to buffer the variability of demand and supply lead times. The focus on outbound logistics is to balance customer service level against cost. Cost of distribution is not just transportation costs but also includes warehousing including special requirements such as refrigeration, insurance and financing of stock, and stock slippage (deterioration, damage, pilfering and obsolescence). The more stock that is held the greater the cost of storage and the greater the chances of losses.

The main components of distribution management are:

- Distribution strategy
- Warehouse operations
- Stock management
- Transport planning
- Partnership with customers

Strategy

It is important that a company in a consumer focused business has a defined distribution strategy. The first criterion of the strategy is to decide whether the management of activities should be by the company or by a third party. With assets (buildings, equipment, transport vehicles) the strategy can go three ways: own the assets or some of the assets, lease or rent assets, or use contractors. Some of the various strategy mixes are shown in Table 4.1. Note there are 64 possible combinations, e.g. own premises, leased premises, own management of premises, third party management of premises, own transport, leased transport, or third party supplied and managed transport, and so on.

Table 4.1

	Warehousing		Transport	
	Building	Management operation	Trunking	Management delivery
Strategy A	Own	Own	Own	Own
Strategy B	Rented	Own	Leased	Own
Strategy C	Rented	Own	Third party	Third party
Strategy D	Own	Own	Third party	Third party
Strategy E	Rented	Third party	Third party	Third party
Strategy F	Rented	Own	Own	Own
Strategy G	Own	Own	Third party	Own

There are some obvious advantages of distribution management by a third party, e.g. the distribution expertise of third party companies, and the avoidance of capital outlay and under utilized equipment. However, as the delivery of the finished products is closest to the customer on the supply chain, there could be some degree of risk if the management of outbound logistics is totally left to third parties.

It is important that, for a manufacturer of fast-moving consumer goods (FMCGs), the distribution strategy should consider the opportunities for both present and future business through an appropriate mix

42 TOTAL MANUFACTURING SOLUTIONS

of the channels of distribution, e.g. supermarkets, wholesalers, and direct to retailers. The distribution strategy should also include the company policy of exclusive agents or stockists and of direct mail order to consumers. Figure 4.5 illustrates an example of the channels of distribution in a typical FMCG business.

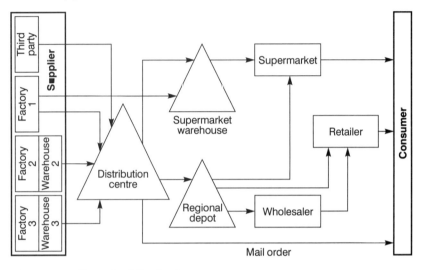

Figure 4.5 *Channels of distribution*

The selection of a strategy may be influenced by the cost of distribution and it should be tempered by the business judgement of customer service and future opportunities.

The location, design and operations of distribution warehouses are all vital ingredients of a supply chain – not only for cost optimization but also for the quality and safety standards of products and for improving customer service by a faster turnaround at the warehouse. There are computer simulation models available for determining the size and location of a distribution centre, but local body planning regulations, the proximity of a highway and a big demand centre very often will be the prime determinants of the location.

Warehouse operation

The operations of a distribution warehouse in general, can be represented by Figure 4.6. There are good opportunities of 're-engineering' the warehouse functions when the total process from reception to despatch is critically examined.

The design issues of a warehouse include:

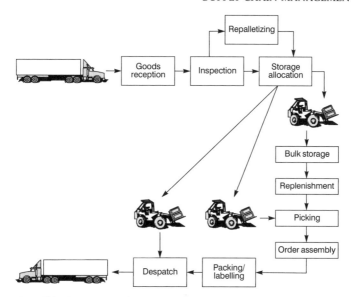

Figure 4.6 *Warehouse operations*

(a) **Storage systems**
- block stock
- back-to-back racking
- double deep racking
- narrow aisle racking
- drive-through racking
- mobile racking

(b) **Handling systems**
- counterbalanced trucks
- reach trucks
- turret trucks
- stacker cranes
- automated guided vehicles
- overhead cranes

(c) **Product quality**
- ambient
- chilled store (e.g. margarine)
- cold store (e.g. ice cream)

(d) **Safety and control systems**
- detection systems
- sprinkler and fire hydrants
- warehouse management system software

Stock management

As indicated earlier, stocks are kept as a buffer along the supply chain in various warehouses, factories (work in process) and retail store shelves. These inventories can cost between a minimum of 15 per cent up to 40 per cent of their value per year (storage space, handling costs, energy costs including heating and refrigeration, stock slippage and insurance). Therefore careful management of stock levels makes good business sense.

In traditional stock management there are two basic approaches: the pull approach and the push approach. In a pull system (Figure 4.7) a warehouse is viewed as independent of the supply chain and inventory is replenished with order sizes based on a predetermined stock level for each warehouse. The stock management model for the pull system is normally geared to establish ROL (re-order level) and ROQ (re-order quantity). That is, when the stock drops to a certain level, a re-order is triggered of a predetermined amount. The re-order quantity takes into account past demands and the lead times for a re-order to be satisfied. The aim is to have as small amount of inventory as possible on hand at any one time, and the reorder quantity should likewise be as small as possible. However in some processes, such as a batch system, there will be a minimum amount that can be produced and in other cases there can be economies of scale which will determine the optimal size of an order. The push method is used when economies of scale in procurement outweigh the benefits of minimum inventory levels as achieved in the pull method (Figure 4.8). That is the warehouse does not decide the quantity of the order but will receive a delivery as determined by the production schedule. Normally a fixed interval review model with a forecast

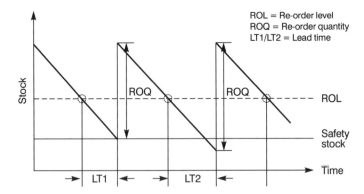

Figure 4.7 *A basic ROL/ROQ model for a 'pull' system.*

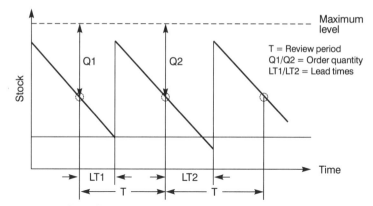

Figure 4.8 *A basic fixed interval model for a 'push' system*

demand for manufacturing planning is used in a push system.

With the support of information technology, businesses are moving towards a virtual inventory system with a single stock concept which can be held anywhere in the system, be it on order with the supplier, in production or at the point of sale. This is the concept of virtual inventory management (VIM) or electronic inventory. Thus instead of considering stocks of raw materials, work in progress at the various stages of production and finished goods in warehouses each as separate stocks of inventory, purely because of their physical location, inventory is now considered as being part of one single stock.

Transport planning

Transport planning is a key decision area of distribution management. Transportation is a non-value-added item to the cost of the product and absorbs, in general, the biggest share of the logistics cost. Students often argue that unless a product is in the right place it is of little value and thus transportation does add value. Not so! The concept of adding value relates to the transformation process, that is the conversion of inputs of raw materials, labour and machinery into a finished product. Storage, inspection, and transportation all add cost but do not add value. Some of these costs will be unavoidable; materials have to be moved, goods have to be distributed, but storage, handling and movement only add to the cost, and not to the value of the product.

The main factors in transport decisions are:

- Transport mode selection
- Trucking routing
- Delivery planning

There are various means of transportation such as railway, river, canal, coastal shipping and pipelines for products such as oil. In some countries, for some products, air transport might prove to be the most viable option. Generally however because of dependability, flexibility, speed and door-to-door service, road transport has proved to be the best option. For Great Britain the Channel Tunnel has added to the convenience of road transport.

There are significant opportunities in optimizing the selection of hauliers or type of trucks. In order to take advantage of the competitiveness and the up-to-date development of vehicles, companies are building partnerships with hauliers.

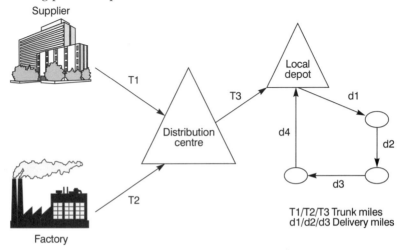

Figure 4.9 *Distribution routes*

After the selection of the mode, the planning of trunking or primary transport for single-drop repetitive journeys between known or well known locations (e.g. factory to warehouse), is relatively straightforward. However the routing and scheduling of delivery vehicles to customers is extremely variable and therefore requires more systematic planning. There are computer-based procedures to optimize delivery to customers. The objective is not to minimize the total mileage but to maximize the utilization of vehicle time (delivery window) and space (by volume or weight) and ensuring customer service.

Partnership with customers

The partnership with customers is the mirror of working with suppliers with the role reversed. Ideally the relationship will be that the customer involves the manufacturer in the market research phase so that together the best product can be designed to meet the end consumers' needs. Likewise the customer through electronic data information (EDI) can input directly into the MRP II system. Improved internal relationships within the business between manufacturing and logistics staff interfacing directly with the customers should achieve a more precise specification of customer needs and sharing data (e.g. EDI). This is the ideal situation. However it is not practicable to form true partnerships with all customers.

Thus it is useful to carry out an ABC analysis (Pareto chart) to identify the top customers as shown in Figure 4.10. The Pareto theory is that 20 per cent of the customers will account for roughly 80 per cent of the business. ABC analysis takes this a step further by dividing customers into three groupings as shown Figure 4.10. Normally the division will be the top 5 per cent, the next 15 per cent and the balance of customers – 80 per cent. In this example the analysis has been further broken down so that it can be seen that the top five customers account for 24 per cent of the sales, and overall just 3 per cent of the customers account for 80 per cent of the sales.

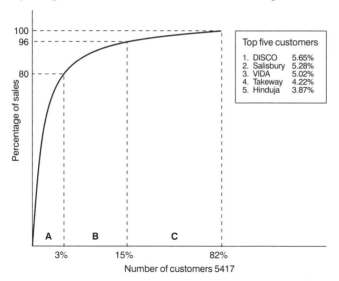

Figure 4.10 *Activity-based costing analysis – customers*

Another challenge of working with customers is to identify the true profitability of all customers and then to improve the profitability of key customers. Figure 4.11 illustrates that a 'tail' of unprofitable customers actually reduces the total profit contributions.

In one organization we encountered, the top 5 per cent of the customers accounted for 40 per cent of the sales, and because of their

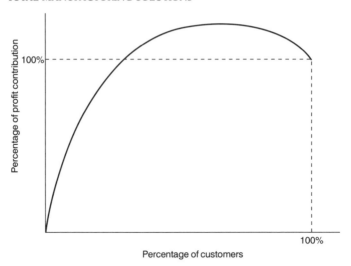

Figure 4.11 *Customer profitability. Source: Christopher, 1992*

importance to the company they had been able to negotiate volume discounts and special delivery arrangements. When these benefits were examined and costed out it was found that whereas the balance of the customers were providing the company with a true 40 per cent gross profit margin on sales, these top 5 per cent were only providing the company with a margin of 10 per cent. Thus overall the gross margin on all sales for the company was reduced to 28 per cent whereas the budget had allowed for 40 per cent. This had not been apparent as the discounts had been shown in the accounts as an overhead expense and the extra transport costs had also been included as an overhead cost. There were also other reasons as to why the drop in true margin was not obvious.

In order to assess the true profitability of customers it is necessary to move away from the average allocation of cost (e.g. cost per tonne) and conventional cost accounting. Logistics managers are now working towards what is known as 'activity-based costing' (ABC) where cost is allocated according to the level of activity that consumes the resources. For example, the order picking cost of an order will vary according to its work content depending on whether the order is in full pallet or small units, number of lines or SKUs or whether it requires additional packaging.

Foundation 6: Supply-chain performance

We have identified supply chain performance as an important foundation stone of total manufacturing solutions because:

- we need to have a grip over the various parameters of supply chain in their true perspective, and
- the performance measures can be used not only to drive continuous improvement of the business process but also to set directions for future strategy.

Therefore the criteria for performance measures should cover a balanced approach to all key parameters of the supply chain and should provide operational measures rather than financial measures. Measures should be simple, easy to define and easy to monitor.

In determining what should be measured it is useful to get away from standard accounting measures. Operations requirements are different from the accountants. In determining our own measurements we should ask:

- What should be measured and why?
- What is the benefit of this measure, how does it help us to achieve our goal?

Once we decide what should be measured, then we can determine how it should be measured.

Measurements are only of any use if they are fed down to the workers and if the workers understand what the measurements mean. Ideally if a worker receives a measurement then that worker should be encouraged to become involved in finding ways to improve the system so as to achieve improved results.

Measurements should never be used as a means of levelling blame to one department or to criticize any one individual. Measurements should be aimed at finding where problems occur so that action can be taken so as to prevent future mistakes. After all no one section or department works alone, we are all in this together. If the company goes down we all go down!

The methods given in Figure 4.12 are fairly standard and should fit into most systems without upsetting the accountants. None the less, no measurement is sacrosanct and each measurement should be challenged. A measurement that doesn't help to improve the system is an unnecessary cost.

Planning performance

'Stock turn' is the ratio of the total sales (or throughput of a product) and the actual stock at any time, both being expressed in either money or volume. The objective is to maximize the stock turn (i.e. minimize average stock level) but also to maintain stock availability. Stock availability

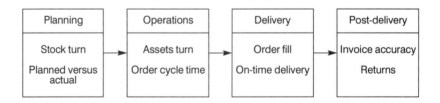

Figure 4.12 *Supply-chain performance*

(the percentage of demand that can be met from available stock) is another measure of performance; availability can also be measured by the number or percentage of orders satisfied within a given target time frame.

The unit of stock turn is a number or ratio. It is also a common practice to express stock profile in terms of equivalent weeks or days of stock. For example if the cost of goods sold (raw materials plus direct labour and other manufacturing costs but not overheads) is £25 000 and the amount of stock of finished goods on hand totals £5000 then the number of days of finished goods equals 73 days (5000/25 000 x 365 = 73). That is, on past performance it is going to take just on two and half months to sell all the finished goods we have on hand. Assuming that we have already paid the suppliers and have paid our workers' wages and paid the other costs of production, this obviously means that our inventory of finished goods is putting pressure on our cash flow. The same types of calculations can be made for stocks of raw materials and work in progress.

One company we visited was proud of the fact that in their high street stores that they only ever had seven days of retail stock (own product) on hand. Their reorder system to their central warehouse was on-line and re-orders were delivered within 24 hours. The warehouse of finished goods held six months' stock, and the stockpile of raw materials for production amounted to seven months' supply. Assuming suppliers were paid within one month of supply this meant that this company was waiting twelve months and seven days to recover the cost out-laid for stock! Not really anything to be proud of when looked at in this fashion.

The share of stock by primary materials (i.e. raw materials and packaging materials), work in progress and finished products varies according to the products and method of manufacturing as illustrated in Figure 4.13. 'Planned versus actual' (also known as planning efficiency) is a simple measure of whether the plan is being achieved. This measure can be for any period, i.e. this month we planned to produce 80 000 units, but our actual production was 70 000 units. Therefore we were 87.5 per cent efficient. This measure is of little use if we cannot trace back to why pro-

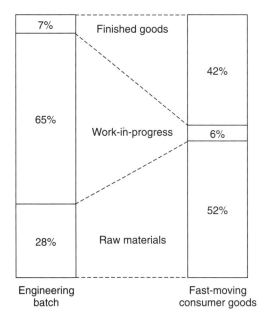

Figure 4.13 *Stock profile: percentage of total stock*

duction was short of the plan, not with a view to criticize but with a view of correcting the system so that we will be more efficient in future. It is more meaningful when planning efficiency is expressed for each product or SKU rather than for total volume. Sometimes this measurement will be more hard hitting if it is expressed in lost sales.

It is a common practice to express both the planned and actual production of each week in graphs and calculate planning efficiency figures for the week and cumulative year to date. But all this effort is only of any use if the information, however expressed, leads to corrective action being taken. Too many measures too often will only serve to confuse the real issues.

Scott and Westbrook (1991) introduced an apparently simple tool, called a *pipeline map*, to present a snapshot of the total stock in a supply chain. As shown in Figure 4.14, the supply chain of an FMCG product is mapped by a series of horizontal lines representing the average time spent in major processes between stock holding points, and a series of vertical lines showing in the same scale (e.g. days) the average stock cover at each point. Pipeline volume is the sum of both the horizontal and vertical lines and represents the time needed to 'flush' the inventory in the supply chain at an average rate of throughput. Pipeline mapping is a useful tool to understand the planning performance of a supply chain, but additional analytical techniques should be used to identify the key areas of improvement.

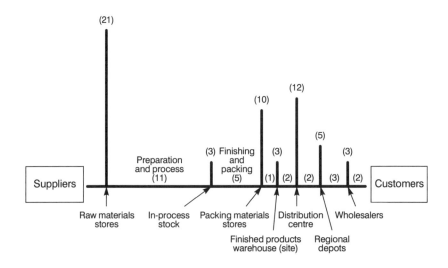

Figure 4.14 *Pipeline map of an FMCG product*

Operations performance

'Asset turn' is the ratio of total sales and fixed assets. It is important that the value of fixed assets is updated by taking into account the depreciation rate for the type of asset according to a defined accounting policy of the company. Assets utilization (time based) is more relevant to all manufacturing performance. However the measure of assets turn (value based) provides an indication of investment in the supply chain. In the short to medium term this measurement is of little use as the investment in the assets has already been made and the measurement is against a past decision. In biblical terms the sins of the fathers are being visited on the next generation.

'Order cycle time' (also known as lead time) is the elapsed time from the placement of an order by the customer to the receiving of delivery (see Figure 4.15). It is important to state standards to suit customer requirements and analyse the total cycle time into relevant components.

Lynch and Cross (1991) claim that only 5 per cent of cycle time is devoted to adding value. In many cases the product is waiting to be worked on 95 per cent of the time. (This excludes raw materials in stock and finished goods in the warehouse.)

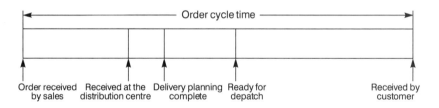

Figure 4.15 *Order cycle time*

Delivery performance

'Order fill' is the percentage of first time satisfied orders. From the customer's point of view this is probably the most important measurement. The order is the correct quantity and quality. The next most important measure as far as the customer is concerned is if the delivery is on time!

'On time delivery' can be expressed as a percentage of full orders delivered on time. 'On time' may be determined by the standards of order cycle time set for a customer or the agreed date of delivery as set by the customer.

Past delivery performance

The times a customer normally comes into contact with the company will be when they are placing an order, that is when the sale is made, when they receive the goods, and when they pay for them. These are the three crucial times for the customer. These are the times that your customer will judge you. Whatever happens in between in the manufacturing process, the supply management or whatever is not of any great interest to the customer.

Payment will generally be against your invoice. It follows that if the invoice is incorrect – that is wrong quantities shown, incorrect prices, and discounts not allowed – that the customer will be irritated. An incorrect invoice also gives a customer an excuse to delay payment. From our experience few customers need any excuse to delay payment.

With invoicing time is of the essence. Most organizations have a cut-off date for passing invoices for payment. If the cut-off date is the twelfth of the month and your invoice is not received until the fifteenth, then for the sake of three days you are going to wait 30 extra days for payment.

If your company is having problems with a large number of incorrect invoices, then measurement of the number of invoices wrong as a percentage of total invoices could well prove a worthwhile measure. It will not be difficult to determine if there is a problem. Five minutes with the accounts receivables' clerks over a cup of coffee should be sufficient.

Likewise a target date should be set for despatch of invoices. We can see no good reason as to why invoices cannot be mailed out within 24 hours of delivery of the goods. With an integrated system the release of goods for delivery should automatically trigger an invoice.

Return of goods is another key measure. Goods will be returned because they are not what was ordered, or because they are defective, or delivery is late. Our motto should be 'never give the customer a reason to leave'. Delivering faulty goods, or the wrong goods, or delivering late is giving the customer the perfect reason to leave.

Measurement of returns can be expressed as the number returned, or the value of returned goods or as a percentage of deliveries. The value of goods returned is not the true cost. The cost includes the cost of putting right and the loss of confidence experienced by the customer.

Any error can be measured in monetary terms or as a percentage. And don't forget the cost of putting things right. However what can't be measured is equally important.

What can't be readily measured is loss of confidence by the customer, loss of future orders, loss of reputation with possible new customers, loss of morale amongst workers, an acceptance that being right 90 per cent of the time is good enough. Imagine what would happen at Heathrow if the air traffic controllers were only right 98 per cent of the time, or if 5 per cent of babies were dropped by the doctor at delivery? Anything but a target of 100 per cent correctness is telling workers that we expect that there will always be some mistakes and so therefore mistakes are acceptable.

Customer service

As Christopher (1992) says, 'Customer service is one of the most powerful elements available to the organization in its search for competitive advantage and yet is often least well managed'.

The output and performance of all supply-chain management activity are relevant to customer service. However, the indices of both delivery performance and past delivery performance are directly related to customer service. It is useful to monitor the indices separately. It is also possible to calculate a 'composite index' by attaching a weight to each service index. The weighting should reflect the relative importance that the customers attach to these elements.

Table 4.2 as shown by Christopher (1992) illustrates an example of composite customer service index.

Table 4.2

Service index	Weighting (%) (a)	Performance (%) (b)	Weighted score a x b
Order fill	45	70	0.315
On time	35	80	0.28
Invoice accuracy	10	90	0.09
(100 – returns)	10	95	0.095
			.78

Composite customer service is 78 per cent

No matter how good our internal measurements are, the best way of finding out how good we are at customer service is to ask the customer what is important to them, and then to take action to achieve what the customer wants. It could be, for example, that the customer can never get through on the telephone. We could therefore set a target that no telephone should ever ring more than three times before it is answered and that no matter who is passing they should pick up the telephone and at least take a message. To stress the point that we should always be seen as easy to communicate with, we could also set a target that all faxes are to be answered within 24 hours, and so on. These targets do not have to be measured. It will be a matter of changing the culture of the company so that such actions become standard practice; 'this is the way we do things around here'. No manner of measures will ever make up for a strong culture driven by pride in achievement and getting things right first time and every time.

Summary

We began this chapter with a quote from Proust, and the importance of 'seeing with new eyes'. We continue from where we left off in Chapter 3, with the need for measurement and for a customer focus. We extend this approach to include a supplier focus.

In the past the attitude in business was secrecy and restricting information to a 'need to know' basis. This especially applied to suppliers, where the aim was to achieve the best deal possible.

Supply-chain management asks us to see with new eyes: to see customers and suppliers as part, and indeed partners, in our overall production process. As partners both suppliers and customers can be, and should be, involved in the planning process.

A major tool in achieving this is the computer-driven MRP II system which is discussed in detail in this chapter. Other key areas discussed in this chapter are distribution strategies, materials handling, warehouse management and stock management. Our main lesson in this chapter is to open our eyes, to appreciate how much we can benefit through honesty and loyalty to suppliers and to customers.

References

Ballon, R.H. (1992) *Business Logistics Management.* Prentice Hall International.
Christopher, M. (1992) *Logistics and Supply Chain Management.* Pitman Publishing.
Hammer, M. and Champy, J. (1993) *Reengineering the Corporation.* Harper Business.
Homlihan, J. (1987) *Exploiting the Industrial Supply Chain, Manufacturing Issues.* Booz Allen & Hamilton.
Lynch, R. L. and Cross, K.F. (1991) *Measure Up! Yardsticks for Continuous Improvement.* Blackwell.
Scott, C. and Westbrook, R. (1991) New Strategic Tools for Supply Chain Management. *International Journal of Physical Distribution and Logistics Management*, 21, No. 1. MCB University Press.

5
Environment and safety

> *The River Rhine, it is well known,*
> *Doth wash your city of Cologne;*
> *But tell me, Nymphs, what power divine*
> *Shall henceforth wash the river Rhine?*
>   S.T. Coleridge

Environment and safety issues are not popular issues with manufacturers. But none the less they are important issues and should be considered as critical success factors. If they are not considered important for moral or ethical reasons then certainly they should be considered important in terms of money.

Safety, or rather lack of safety, in the product or in the factory will inevitably cost money. Accidents mean lost production time plus tiresome time wasting inspections by government officials and perhaps legal costs as well as the cost of correcting the situation so that the accident will not occur again.

It is said that if an accident can happen then it will, and usually at the worst possible time! Prevention will be always cheaper than putting right.

Since Wickham Skinner of the Harvard Business School published his famous article, Manufacturing – the Missing Link in Corporate Strategy, in 1969, hundreds of books and articles have been written on manufacturing strategy, mission or policy for industries. However few of these publications describe the importance of environment and safety issues as an integral part of manufacturing policy or process. The references that there are seem to emphasize either a 'structuralist' philosophy where the success in a manufacturing business is believed to come from vision, planning, measuring and monitoring the performance – or a 'behaviourist' approach where it is believed that change is achieved by motivation, involvement, learning and empowerment. Other than meeting legal requirements, environment and safety issues are not often included as critical success factors.

Not surprisingly, there are instances in recent history where the performance of manufacturing businesses was drastically affected due to negligence in environment and safety standards. A failure in product safety which caused deformed 'thalidomide children' is still haunting the manufacturers. The gas explosion of 1984 in Bhopal, India, which killed over 1000 people, permanently damaged the business of the manufacturers. Food poisoning costs to John Farley and Wests were huge. Environmental pollution by chemical companies in New Jersey resulted in numerous legal battles with consumers and also affected their business performances. On a global scale industrial pollution is the main contributor to the so called 'greenhouse' effect.

Environment and safety are not just social or political issues, they are vital ingredients contributing to the performance of a manufacturing operation. In particular, in manufacturing industries, there is much scope for environment and safety accidents to occur and likewise there are many opportunities to prevent accidents. Where operations are repetitive in nature, common environment and safety factors can readily be identified and controlled.

We have categorized the pillar of 'environment and safety' into three 'foundation stones' as follows:

7. Product safety
8. Industrial safety
9. Environmental protection

Foundation 7: Product safety

Too often do we see advertisements requesting drivers of a particular model to return their car for some safety modification. What sort of message does this give the marketplace? Have you ever wondered how much the cost of the recall really is, not only the cost of advertising, but the cost of putting the product right, and of course the lost opportunity costs. The waste of time, effort, and money can usually not be fully quantified and when an organization does go through the trauma of a product recall or even a withdrawal, direct money costs may well be the least of the problems.

A major incident in product safety especially if related to consumable and brand products can seriously damage the brand and also work against the goodwill and business of the whole company. This is particularly obvious in food products. If all the faulty products are not quickly identified and recalled the result could be extremely costly both in terms of health and money. If the faulty products contain toxic materials, they could be absorbed either through the respiratory system, or through the skin or through the gastro-intestinal tracts. If micro-organisms are present in a product their fast propagation rate will spoil the quality of the product. Some micro-organisms may produce toxic substances and others may be harmless to health but they can have adverse effect on the quality of consumable products. A concentration of ten million bacteria per gram of a food product generally means that the product is spoiled. Only one micro-organism in water can generate seven hundred million micro-organisms in just 12 hours. The physical damage of a product down the supply chain may be costly but less harmful than a chemical damage.

Cost of product safety

In addition to its effect on goodwill and health, the financial cost related to a product safety incident is generally compounded by costs at various levels such as:

- Cost of investigation
- Cost of sterilization of the plant and storage areas
- Cost of production downtime as the plant is shut down until a certified clearance is obtained
- Cost of recall from the customers and consumers
- Cost of salvage if the returns are worth salvaging through reprocessing or repackaging
- Cost of dumping if the returns could cause health and environment problems
- Legal costs and damages

And of course it is not only your immediate customers who will be dissatisfied; withdrawals and recalls of products, no matter how well handled, will be well publicized by the media to the whole marketplace – potential customers and competitors alike.

Control of product safety hazards

The safety hazards of consumable manufactured products are manifested in various forms including:

- **Toxicological**   The presence of poisonous or toxic substances in a product or process.
- **Micro-biological**   The presence of micro-organisms or bacteria in a product, plant or process.

60 TOTAL MANUFACTURING SOLUTIONS

- **Chemical** Corrosive actions or chemical reaction with other agents and decomposition under heat.
- **Physical** Damage to the product and by the product or its packaging at any stage during manufacture, storage or distribution.
- **Environmental** Pollution of the environment by the product and its packaging materials especially if they are not bio-degradable.

In a manufacturing and supply chain process, where activities are repetitive, the above hazards can be prevented by measures as illustrated in Figure 5.1

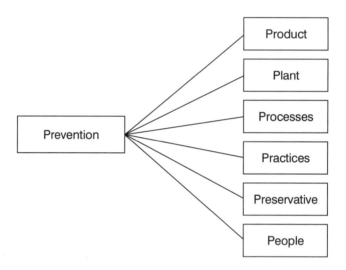

Figure 5.1 *Product safety measures*

Development and design engineers must avoid features in product design that enhance the probability of errors or safety hazards at the subsequent stages of the supply chain. A formal safety clearance of all new products regarding toxicological hazards must be mandatory.

Plant and equipment for the processing and packaging of consumer products, especially edible products, must follow the hygiene design aspects with regard to ease of cleaning and micro-biological impermeability. The paper, Hygiene, Design and Operation of Food Plants, by E.R. Jowitt and E. Horwood (1989) is recommended to those engaged in the operation of a food plant.

A clear understanding of the sources of contamination is vital to ensure product safety in a manufacturing process. Raw materials can be controlled by micro-biological tests before the start of the process. Water is the main source of micro-organisms, and a water decontamination system (such as chlorination, pasteurization) is an essential part of a process containing water. A process with a cleaning-in-place (CIP) system is also recommended.

The code of practice at each stage of the supply chain including raw material approval, sanitation, house keeping, storage and handling and fault precautions, and the strict adherence to them is the preferred practice of product safety. Many companies have formalized the quality assurance procedures through ISO 9000 certification.

Preservatives are an essential ingredient of a consumable product to prevent accidental or normal level of contamination. However, the reliance on preservatives is a protection against a low level of product safety.

People are the key to success to any programme, including one for product safety. The company's organizational structure should include a quality assurance manager qualified in microbiology to ensure the quality standards of the product. This person's team should be supported by a multi-functional team from the factory and a continuous training programme to ensure the safety and quality standards of products.

The hazards and preventive measures of product safety at different stages of the supply chain are summarized as illustrated in Figure 5.2.

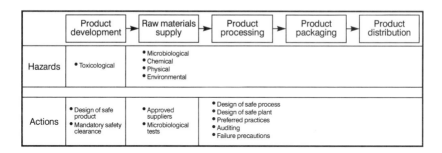

Figure 5.2

Foundation 8: Industrial safety

Since the days of the Industrial Revolution safety standards at the workplace have significantly improved. However, processes and equipment have attained a high degree of complexity. We are now dealing with over a million chemicals and compounds with a massive reserve of energy in process. Unfortunately, many managers assume that accidents are part of doing business and that costs are borne by insurance companies. It is important to understand that the same factors which are creating accidents are also creating losses in production, quality, sales and profit. It is not surprising that the most respected organization specializing in industrial safety is called the International Loss Control Institute (ILCI).

The major industrial accidents in recent times, including Three Mile Island, Mexico LPG fire, Bhopal and Chernobyl, were all well publicized because of their high fatality and serious injury rates. However, there are many more minor accidents that are taking place in manufacturing industries – as demonstrated by a study of 1.75 million industrial accidents in 1969 by F.E. Bird of the ILCI. This is known as the 1-10-30-600 ratio, as illustrated in Figure 5.3.

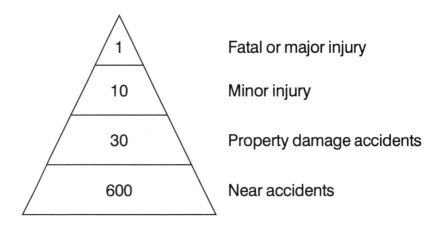

Figure 5.3 *Accident ratio triangle*

The injury or illness costs, which are usually claimed from an insurance company are a relatively small part of the total cost as shown in the 'accident cost iceberg', as illustrated in Figure 5.4.

Causes of accidents

According to the ILCI, an accident is usually the result of contact with a substance or a source of energy above the threshold limit of the body or structure. Dominoes have been used by ILCI to illustrate the causes and prevention of accidents (Figure 5.5).

ENVIRONMENT AND SAFETY   63

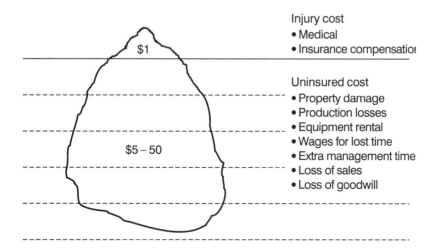

Figure 5.4 *Accident cost iceberg. Source: Bird and Germain, 1990*

The result of an accident includes losses to people, property or process. An incident is the event that precedes the loss; it is the contact with energy or substance. The immediate causes of accidents are the circumstances that immediately precede the contact. They are usually sub-standard acts and unsafe conditions. Basic causes are the real reasons why sub-standard acts or conditions occurred. There are two categories of basic causes, either personal factors (e.g. lack of skill, stress, laziness, taking short cuts, and stupidity) and job factors or work environment (e.g. poor engineering, no safety guards, poor training, lack of maintenance of equipment, etc.).

There are three common causes for lack of management control:

- an inadequate programme,
- inadequate programme standards, and
- inadequate compliance with standards.

In a more practical sense, major factors contributing to accidents in a factory (i.e. basic causes relating to job factors) are:

- Material hazards
- Process hazards
- Mechanical/electrical equipment
- Working practices

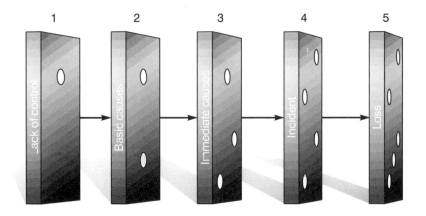

Figure 5.5 *Causes of accidents*

The hazards for frequently used chemicals such as ammonia (toxic), hydrogen (explosive) and hexane (highly inflammable), must be taken into account in the handling, storage and processing of these materials.

The hazards of mechanical and electrical equipment include unguarded moving equipment, pressure vessels and electric shock.

Lack of protective wear, inadequate training, improper tools and procedures are examples of working practices conducive to accidents.

Prevention and control

The three stages of accident prevention and control are:

- pre-contact,
- contact, and
- poor contact.

Pre-contact control is aimed at the prevention of accidents. The leading manufacturing companies of the UK are increasingly commissioning HAZOP (HAZards and OPerability) studies to identify potential hazards both before and after the installation of a manufacturing plant or process.

Contact control measures are applied to minimize the amount of energy exchange or harmful contact. The ILCI guidelines for contact control include:

- Substitution of alternate energy forms (e.g. electric motors to replace shafts and butts).
- Reducing the amount of energy used (e.g. low voltage equipment, or use of transformers).
- Placing barriers between the source of energy and people (e.g. fire walls, guards, fire fighting equipment).
- Modifying contact surfaces (e.g. padding points of contact).

Part contact control applies to actions after the accident such as the implementation of emergency action plans. This does not prevent the accident but minimizes the damage.

Fire fighting equipment, systems and procedures should embrace all three stages of accident prevention. Depending on the size and type of operation, the key fire fighting features should cover:

Equipment
- Hydrant system
- Water storage
- Extinguishers

Systems
- Sprinklers
- Smoke detectors
- Fire alarms
- Procedures
- Fire fighting organization
- Fire escape markings
- Fire fighting drill

Implementation and monitoring

Each factory should deploy a properly trained safety manager to implement and monitor safety programmes. A continuous safety training programme for everyone in the factory should be in place. The management should promote safety awareness by encouraging safety posters, safety weeks and safety awards.

It is important to measure and monitor the accident rate of the factory. The most widely used index is measured as illustrated:

$$\text{Accident rate} = \frac{\text{Number of accidents} \times 200\,000}{\text{Employee's hours of exposure}}$$

It is encouraging to note that there has been a gradual reduction of accident rates in most of the industrialized countries.

For an individual organization a formal audit by accredited safety auditors is a sensible way to assess that safety standards are being met and to learn of improvements that can be made. Many leading companies follow ISRS (International Safety Rating Systems). Another recognized standard is the HSG (Health and Safety Guidelines).

Foundation 9: Environmental protection

Environment protection is an important international issue and industrialized countries are spending between 0.5 per cent and 1.5 per cent of their gross national product (GNP) on the control of pollution. It is a big subject and any attempt to make a comprehensive analysis of all the issues is beyond the scope of this book. Our emphasis is focused on the impact of environment in manufacturing companies. The Advanced Studies Centre of the Massachusetts Institute of Technology back in 1976 studied the cause and effect of environment factors on the performance of a wide range of companies from different industrial sectors. It found in all cases that those companies which were most advanced in environment protection were also the most profitable. On reflection it is not surprising that an efficient (and profitable) company will be safety conscious and environmentally aware and will be following best practices. It is however surprising that the investments for environment protection by manufacturing companies swing to the political pendulum rather than to business objective. Environment protection is going in cycles without showing a continuous improvement. In the 1970s the environment was a political hot potato but as we became accustomed to the issues, and without doubt some issues were overstated, interest tended to wane. But now during the 1990s, influenced by pressure groups such as Friends of the Earth, and well publicized activities of Greenpeace, environment issues are again at the forefront. If you have any doubts ask Shell Oil.

Environment protection relates to pollution control in two stages. Conventional controls or 'first generation pollution' controls are applied to pollution in air, water and of noise created in the manufacturing process. Such controls are usually regulated by legislation. There is also a 'second generation pollution' which relates to the problems caused by the usage of certain products and chemicals over a long period. The most widespread example of such 'second generation pollution' is the contamination of land which permeates ground water.

Causes of pollution

Pollution control engineering has essentially evolved from sanitary engineering and thus the solutions are primarily concerned with effects rather than causes, and with control rather than prevention. The overall ongoing economic impact of pollution has been largely neglected and most of the attention of manufacturing companies has gone to the cost impact of pollution control.

The contamination of land is mostly caused by the disposal of solid wastes by manufacturing industries. With the introduction in the UK of the 'land fill tax' the disposal of solid wastes by incineration will be more cost effective and environmentally friendly in the future.

The three main gases causing air pollution are carbon dioxide, sulphur dioxide and nitrogen oxides. For many years the consumption of combustion fossil fuels has been releasing carbon dioxide to the atmosphere faster than it can naturally be absorbed by photosynthesis (provided by trees and plants). As the proportion of CO_2 in the air increases, it absorbs heat and as a result the atmosphere warms up. Sulphur dioxide resulting from the combustion of coal and oil or any sulphur burning process is another pollutant of air and one of the substances causing 'acid rain'. The damage by acid rain to plants and trees is very evident in parts of Europe. Other acidic gases are the oxides of nitrogen resulting from high temperature combustion processes in power plants.

Lead is a serious pollutant (neurotoxin) affecting nerves and brain. The sources of lead include emission from motor vehicles, lead pipes carrying drinking water, paint and other industrial processes. The Royal Commission on Environmental Pollution recommended in 1983 the benefits of banning the use of lead in petrol. A second pollution bearing metal is cadmium which is used industrially in batteries, metal plating and micro electronics. The discharge of cadmium from local industries in the Severn Estuary in the UK severely damaged the local shell fish industry. A third heavy metal is mercury, causing hazards to life even today. In the 1950s the discharge of industrial effluents with high levels of mercury in a Japanese bay led to deformity and death for villagers who ate the fish from the bay.

Another harmful mineral is asbestos, causing painful and fatal diseases such as asbestosis and mesothelioma. Many domestic items such as textured ceiling, ovens, electrical heating equipment in the past contained asbestos. After campaigning by environmental pressure groups, asbestos lagging in power stations and electric sub-stations has been gradually eliminated in the UK.

The noise levels in many 'metal bashing' and packaging industries caused low performance and, more seriously, hearing impairment. Today there are established preventive and protective measures of noise control.

Cost of pollution

In addition to the long-term immeasurable damage done to vegetation, birds, animals and human beings by air and water pollution, there are many instances of huge compensation bills paid by polluting industries.

The notorious case of mercury poisoning in Japan referred to above led to damages of over US$50 million (1971 value) being awarded to 700 people who were crippled and to the estates of 200 people who died.

In 1978, as a result of the wreck of the oil tanker *Amoco Cadiz*, 200 000 tons of crude oil was discharged into the English Channel. The French Government presented claims amounting to $2 billion.

In 1992 Cambridge Water Company (UK) were awarded damages of £1 million in compensation for the pollution of land due to tetrachloroethylene by a local leatherworks company.

Benefits of environmental protection

A sound environment protection policy of a company can earn it an extremely marketable environment friendly image leading to higher sales and profitability.

There are also several published examples of 'non-waste technology' where a project of environment control turned out to be a profit earner. One such example is the Dow Chemical Company's $7.2 million project for the re-use of cooling water which produced over 10 per cent return on investment and considerably reduced the pollution of a neighbouring river.

The famous 3P programme (Pollution Prevention Pays) of the 3M company brought about major savings including $2 million from the elimination of hydrocarbon wastes from a reactive costing process. When 3M instigated this programme back in 1974 the approach was to capture and control pollutions and emissions before they could damage the environment. This approach although effective has been changed to a philosophy of prevention rather than containment. The 3P programme now aims to prevent pollution at source by using different materials, changing the process, re-designing the plant and equipment, and through recycling waste.

Another example is a distillery in Scotland. An effluent treatment project for the control of suspended solids and BOD (biological oxygen demand) produced, with the addition of a drying plant, high quality cattle feed.

Environmental strategies

M.G. Royston (1979) suggested an eight-point strategy of environment protection for a manufacturing company.

1. Cut down waste by improving efficiency.
2. Sell wastes to someone else.
3. 'Build on' extra plant to convert wastes into raw materials or products which are valuable to the company or to someone else.
4. Work with self-cleansing and dispersing power of the environment so as to permit maximum discharge or effluent.
5. Negotiate emission standards and subsidies with the authorities and the community.
6. Build the treatment facility needed for residual wastes jointly with another enterprise or the local authority.
7. Build the plant using company staff and know how.
8. Sell the acquired know how to others with the same problem.

References

Bird, F.E. Jr and Germain, G.L. (1990) *Practical Loss Control Leadership.* International Loss Control Institute.

Hopfenbeck, W. (1992) *The Green Manufacturing Revolution: Lessons in Environmental Excellence.* Prentice Hall.

International Loss Control Institute (1991) *Accredited Safety Auditors – Pre Course Reading.* International Loss Control Institute, USA.

Jowitt, E.R. and Horwood, E. (1989) Hygiene Design and Operation of Food Plants. Unpublished paper. Chichester, UK.

Massachusetts Institute of Technology (1976) *National Support for Science and Technology.* MIT, Cambridge, Mass., USA.

Royston, M.G. (1979) *Pollution Prevention Pays.* Pergamon Press, Oxford.

Skinner, W. (1969) Manufacturing – the Missing Link in Corporate Strategy. *Harvard Business Review*, May-June, 136-145.

Wilson, D, (1984), *The Environmental Crisis.* Heinemann Educational Books, London.

6

Manufacturing facilities

Men, my brothers, men the workers, ever reaping something new:
That which they have done but earnest of the things that they shall do.
                                                              Tennyson

Global warming perhaps not only refers to changing weather patterns; it could equally apply to the changing global market.

Globalization – the reduction of tariffs and virtual elimination of national barriers – has exposed management inadequacies that previously were hidden. Previously a company competed in its own country against local competition using like systems and with similar cultures. Inefficiencies and inadequacies were to a large extent hidden. Comparative economic isolation and protection and the advantages of technical expertise no longer exist. Technological advances are constant and are available to all. The only real opportunity for competitive advantage is to do things better and to achieve world class status.

The pillars we have discussed in the last three chapters are 'external' pillars. We now turn to the 'core' pillars of manufacturing. The subject of the present chapter is manufacturing facilities, and Chapters 7 and 8 deal respectively with processes and people.

The challenge of manufacturing facilities is far more complex than cash-flow management, and the parameters are not of the short term nature of labour and software. The outcome of an investment decision for a manufacturing facility is likely to last for 10 to 100 years. Likewise, it normally takes several years of disciplined effort to transform an existing weak manufacturing unit into a strength for the organization. Wickham Skinner (1969) described manufacturing facilities as either a corporate millstone or a competitive weapon depending on the strategy applied and pursued. Steven Wheelwright and Robert Hayes (1985) defined the four stages of manufacturing's strategic role, as illustrated in Table 6.1.

Table 6.1 *Stages in manufacturing's strategic role. Source: Wheelwright and Hayes, 1985*

Stage	Strategy	Role
Stage 1	Minimize manufacturing's negative potential: 'internally neutral'.	Manufacturing is kept flexible and reactive.
Stage 2	Achieve parity with competitors: externally neutral	Capital investment is the primary means for catching up with competition.
Stage 3	Provide credible support to the business strategy: 'internally supportive'.	Longer term manufacturing developments and trends are addressed systematically.
Stage 4	Pursue a manufacturing base competitive advantage: 'externally suportive'.	Long-range programmes are pursued in order to acquire capabilities in advance of needs.

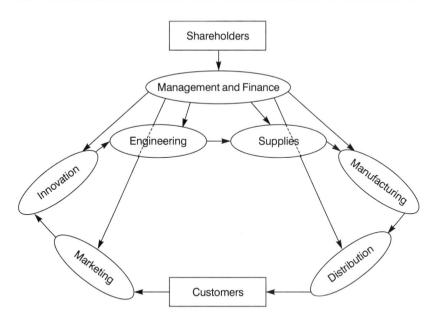

Figure 6.1 *Total business process*

72 TOTAL MANUFACTURING SOLUTIONS

In a manufacturing business, a number of inter-related functions (such as marketing, innovation, engineering, purchasing, manufacturing and distribution) work towards a common objective of satisfying the customers and at the same time ensuring an attractive return on investment for the shareholders. This is illustrated in Figure 6.1. Of these the manufacturing function has the majority share of the company's assets and people. In a typical fast-moving consumer goods (FMCG) manufacturing business:

- 98 per cent of the products sold are either own manufactured or co-produced,
- 90 per cent of the assets of the company are for manufacturing, and
- 75 per cent of the people work in manufacturing.

A simple and popular model is shown in Figure 6.2. This illustrates the manufacturing process where the core conversion activities in a factory are performed by the three key elements of:

- Facilities or 'hardware'
- Procedures or 'software'
- People or 'humanware'

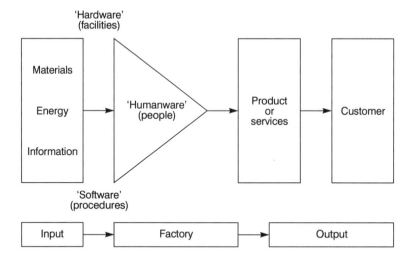

Figure 6.2 *Manufacturing*

It is not enough just to formulate and pursue an 'up front' manufacturing strategy, no matter how good the strategy is. To maintain a competitive advantage it is essential to support the strategic planning of facilities with the ongoing monitoring of performance and with continuous improvement programmes. The management of manufacturing facilities should be dynamic with the relentless pursuit of the elimination of unnecessary non-value-adding expense, and always with the objective of adding value for customers. Competitive advantage once achieved through a strategy such as investment in new facilities will require hard work if the advantage is to be retained.

To achieve and retain competitive advantage it is necessary to consider five elements of manufacturing. We have shown these five elements as the five 'foundation stones' for the core pillar of manufacturing facilities, namely:

10. Sourcing strategy: To ensure a manufacturing advantage in a global market
11. Appropriate technology: To define the optimum choice of processes and an appropriate level of technology
12. Flexible manufacturing: To highlight the ability of facilities to adapt the operation to the changing need of customers, process or materials.
13. Reliable manufacturing: To underline the operational measures for facilities for quality advantage, speed, dependability and efficiency (cost).
14. Manufacturing performance: To define the key indices with which the competitive parameters of manufacturing facilities can be measured, monitored and improved.

Foundation 10: Sourcing strategy

There has been considerable hyperbole regarding world class manufacturing (WCM) and many articles and books have been written on the subject, especially since Richard Schonberger published his book in 1986. There have been a number of interpretations of WCM. Some people associate WCM with working practices influenced by Japan's 'quality movement'. Others understand WCM to be manufacturing at the highest level of performance.

We shall define WCM as the term applied to manufacturers who do achieve dominance in their segment of the global market and who sustain this dominance against world class competition.

Over the last three decades manufacturing strategy has been concentrated on the local area, e.g. for manufacturers in the United Kingdom the concern has been the domestic market and the near neighbours of

Europe. The emphasis has now moved to the determination of either a global strategy or regional strategy, not only for marketing, but for sourcing. Sourcing includes materials and labour, and also includes the basic decision of whether to make or buy.

The globalization of manufacturing began with sourcing and a search for low labour costs. Manufacturing was transferred from the western nations and Japan through the establishment of manufacturing facilities in Asia and Latin America. However it soon transpired that once overseas investment is made in a country the cost of labour creeps up. Additionally as other overseas companies with similar products follow the lead (and move to a country where labour is cheaper) then the initial competitive edge of cheap labour gained by the 'pioneer' company becomes a diminished advantage.

There have been significant changes in the global marketplace, demanding a sound sourcing strategy for the manufacturing company as the changes accelerate. These changes include:

- Newly industrialized countries (the 'little dragons' such as Korea, Taiwan and Malaysia) are acquiring world class manufacturing capabilities. Investors wishing to set up manufacturing in these countries will find labour is not as cheap in real terms as it was 10 years ago. But more importantly these countries are now, in their own right, competitors of world class standing. Other Asian countries which show longer-term potential to achieve WCM in some areas of endeavour include India, Pakistan, Indonesia and of course the once sleeping giant, China. Other regions which could also emerge as WCM contenders include South America and, given time and the resolution of political problems, South Africa.
- The gradual elimination of tariff barriers and the regional pacts for 'common markets' (e.g. Mercosur, NAFTA, Andina, EC, CER, etc.) are encouraging competition from regionally based groupings of countries.
- The emerging markets of what was the 'Communist bloc' are providing new opportunities in the global market. In addition new players are emerging from this block who are close to achieving the status of WCM.
- Improved logistics and electronic communication systems are assisting the implementation of sourcing strategies.
- The growing similarity of what people want to buy across the world is encouraging global product/process development and marketing.
- Investment costs for innovation and new technology are becoming too expensive to concentrate in one local market.

A sound sourcing strategy for a manufacturing company may well be a requirement for the survival of the future. Catching up with the manufacturing performance of the competitors is not enough. The sourcing strategy of the company must move in step with the corporate strategy and reflect, as described in Chapter 3, the marketing strategy and innovation programmes of the company. The sourcing strategy should be dynamic in a relentless pursuit of value to customers in a changing market place. Hamel and Prahalad (1994) analysed the competing requirements of the future and concluded,

> The market a company dominates today is likely to change substantially over the next ten years. There is no such thing as 'sustaining' leadership, it must be regenerated again and again.

In order to develop a sourcing strategy for manufacturing it is necessary to have a formal strategic planning process. The process should be flexible and simple to follow and it should be incorporated with other corporate planning processes. Our strategic planning process for sourcing for manufacturing consists of the eight steps described below.

1. Project brief

The process is best carried through by setting up a project team of about 10 people and defining the brief of the project. The project team should consist of a project director (e.g. head of manufacturing), manufacturing staff (e.g. industrial engineer, plant engineer, manufacturing manger, quality manager), logistics staff (e.g. planning manager, distribution manager), marketing staff (e.g. brand managers), commercial staff (e.g. accountant, buyer and human resources staff).

When preparing the project brief it is useful to have those documents which cover current company activities such as capital investment, annual operating plans and long-term plans. In addition, any other relevant reports (such as information on competition, marketplace, economy and government regulations) of the countries covering the scope of the strategy will help with this activity. The project brief should clearly state the scope, time scale, deliverables and resources required for the project.

2. Manufacturing mission and objectives

The manufacturing mission defines the aim of manufacturing in the corporate strategy or the business plan. The mission statement must fit the capabilities of the manufacturing function. Unless the mission is feasible it will be no more than mere words or rhetoric. Usually the mission statement is described in broad terms as illustrated by the following example:

> The manufacturing mission is to achieve the lowest unit manufacturing cost relative to competition without sacrificing high standards of quality, service and flexibility to the customer.

This mission statement has a priority on low cost. Alternative priorities could include one or more of: quality, customer service, rapid introduction of product, visible presence in emerging markets, combating a dominant competitor, etc. The point to note is that the mission has to be sufficiently specific for a clear objective or objectives to be readily distinguished (see Chapter 11).

Manufacturing objectives consist of performance measures that the company's manufacturing must achieve as part of the annual operating plan. Achievement of the objectives will result in the achievement of the mission.

3. Strategic factors

The understanding and analysis of strategic factors can determine the success of a sourcing strategy. Strategic factors relate to the longer-term implication of both the external and internal factors to project manufacturing into the future. These factors are competition, customer preferences, technology, environment, economic conditions and statutory regulations.

To develop a sourcing strategy for manufacturing, so as to gain a competitive advantage, a detailed competitive position analysis will be necessary. This analysis determines how the strengths and weaknesses of the company's manufacturing position relate to major competitors (both current and potential competitors). The dimensions for this analysis can be cost, quality, dependability, flexibility and innovation. Following this analysis the company should be able to identify any gaps in manufacturing competence and establish priorities for a future strategy so as to gain a competitive advantage.

The strategy of appropriate technology will be discussed later. However it is critical to determine what should be made and whether to make or buy. Such decisions depend on the long-term volumes of the product and of the level of technology required. To identify the preliminary grouping of the sourcing of products, we propose an approach as illustrated in Figure 6.3. This grouping would be finalized after the quantitative evaluation in stage 6.

Systematic analyses of both external factors (such as environment, economic, cultural, social and political factors of the countries

MANUFACTURING FACILITIES 77

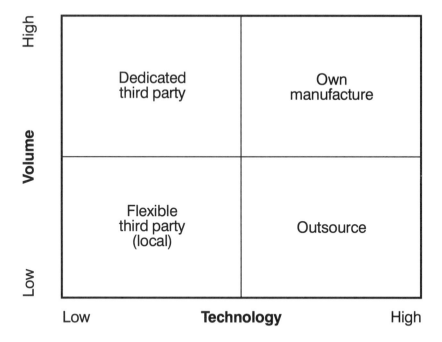

Figure 6.3 *Sourcing strategy*

involved) and internal factors (such as research and development and marketing) should be made to determine how these factors will influence manufacturing.

The project brief may be reviewed and restated after the analysis of all strategic factors.

4. Data collection and data analysis

Once the revised project brief has been finalized, the next stage is the collection of data and the analysis of data. Although the need for data will vary, the following areas will need to be considered:

General information
- Internal information of the company regarding annual plans, long-term plans, R&D and marketing.
- External information regarding competition, economic and political factors of countries involved, social and cultural aspects and environmental (green) issues.

Product information

- Future, 10-year sales forecast by products, and
- Past, 5-year sales history.

Plant information

- Present capacity of own plants.
- Investment plan to increase capacity.
- Present levels of efficiency.
- Other manufacturing alliances, e.g. sub-contractor capabilities.

Stock information

- Stock policy of materials and finished products.
- Warehousing area (space and capacity) and method of storage.
- Method of distribution to customers.

Personnel information

- Projection of people availability and skills.
- Industrial relations of manufacturing sites.
- Amenities required.

Cost information

- Manufacturing costs of products by site.
- Distribution cost elements.
- Cost of warehouse building per square metre.
- Cost of office building per square metre.
- Cost of an employee per year (total cost, i.e. wages, benefits and training).

The purpose of this stage is to calculate the capacity of plant and services for the projected volume and estimate the space required for each activity for each manufacturing site. It is normally sufficient to carry out these analyses for the current year, and at the mid stage and at the completion of the plan or when a significant event (e.g. the manufacture of a new product) occurs. The utilization of assets as determined at this stage should help to establish what to manufacture and where, and the profitability of each site.

5. Strategic options

Strategic options determine how sourcing or own manufacture is going to meet the objectives of the mission. It is useful to reiterate that the objectives refer to performance measures (such as cost, flexibility, quality, etc.) and strategy refers to how these objectives will be achieved.

Strategic options are normally expressed in a number of sourcing scenarios. These are derived from the understanding of the competitive strengths and weakness from the foregoing stages.

As a general rule there should not be more than eight scenarios. Eight scenarios are manageable and enable adequate attention to be given to each scenario. A critical analysis of each scenario is then carried out against the criteria of manufacturing objectives and strategic factors. Two or three scenarios are then shortlisted for quantitative evaluation.

6. Options evaluation

The aim of this stage is to evaluate two or three main options in order to select the best strategy for the future. The analysis should take advantage of simulation modelling tools to select a strategy by optimizing the total operating cost (see Figure 6.4). Costs only need be broad estimates for the evaluation of options.

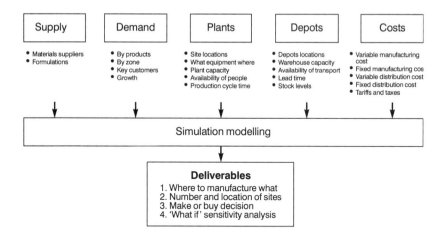

Figure 6.4 *Sourcing strategy model*

The strategy should then be further tested by comparing the investment costs of alternative development plans with quantitative tools such as discounted cash flow (DCF) analysis.

entation plan

...f a sourcing strategy for manufacturing will depend on ...very the changes have been implemented. There should be a structured implementation plan describing the phasing, responsibility, costs and obstacles that have to be overcome.

The strategy itself should not have major changes every year or there will be little chance of maintaining the strategic goal. However, tactics should be continually adjusted to meet changing circumstances.

8. Review

As stated above, there is a need for regular evaluation and review of progress to implement the strategy. In addition to the regular review the entire strategy should be formally reviewed on an annual basis.

Foundation 11: Appropriate technology

Manufacturing technology is popularly seen as the main factor contributing to the manufacturing advantage more than any other elements or foundation stones. However a 'high tech' strategy is not necessarily the best strategy. Selecting or failing to select the correct technology can have long-term and serious repercussions for a business. Technology can give a competitive advantage if managed well. If not managed well, or if the wrong technology is selected then the results can be catastrophic. If an organization is inefficient and not capable of managing standard technology, the introduction of advanced technology will only create further problems.

An international benchmarking survey between the USA, Japan and Europe (Miller et al., 1992) clearly established that 'robotics' ranked as the lowest pay-off of improvement programmes both in the USA and Europe. Technology is responsible for the added-value conversion process. It is a vital foundation stone of total manufacturing solutions, but it is the choice and application of technology appropriate to the product, volume and specifications that create and sustain the advantage. The key issues of appropriate technology are:

- Product and process technology
- Choice of technology
- Evaluation of technology

Product and process technology

Product technology focuses on advanced product innovation to provide the basis for superiority in product performance.

Process technology relates to improvements in the processes of manufacturing a product which is already in existence. The process applies to any added-value conversion operation whether, chemical, metal cutting or packaging.

The benchmarking survey by Miller et al. (1992) indicates that different visions of the factory of the future are emerging in the USA, Japan and Europe. Both Americans and Europeans are emphasizing advanced product technologies while the Japanese companies focus more on process technology and the ability to make rapid design changes in highly customized products. United States manufacturers are moving toward the value factory of the future. European companies are moving toward the borderless factory of the future. In Japan manufacturers are working steadily to build the design factory of the future.

The strategy of product technology must be addressed at the innovation stage (see Chapter 3), but as process technology depends on the product, both process and product development should be carried out as closely as possible. The relationship could be overlapping, parallel or interactive, up stream for the product and down stream for the process. The early commitment of process technology in the design of a product is described as simultaneous or concurrent engineering. The philosophy is to involve participants from marketing, engineering, production, purchasing and quality to work together as a design team to design the process in parallel with the development of the new product.

A 'technological model' (see Figure 6.5) by Brown and Blevins (1989) depicts product technology at a higher plane relative to process technology and the authors argue that the USA has progressively moved from the process technology plane (with emphasis on productivity) to the product technology plane (with emphasis on innovation). This model predicts that Japan is also moving from the strategy of 'wait for a competitor to try a new product' towards the higher plane of product technology. During the 1970s and 1980s Japan's manufacturing productivity increased at an annual rate of double figure percentage and a key factor attributing to this is the extensive application of robotics. R&D expenditure of Japan is constantly rising but it is reported to be 1.8 per cent of sales as compared to 3.4 per cent of sales in the USA. In addition ROI was the first for US companies while market share was the primary consideration in Japan. It is concluded that management philosophies in various companies are related to the local environments of those countries and not necessarily transportable.

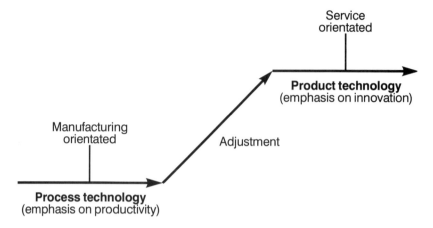

Figure 6.5 *A technological model*

Wheelwright and Hayes (1985), however, concluded that a number of sophisticated high-technology companies which regard product technology as the key to competitive success and process technology are, at best, in Stage 1 of manufacturing (see Table 6.1). They should strive for a longer term competitive advantage to progress towards Stage 4 by pursuing an 'externally supportive' manufacturing strategy including the management of process technology.

Although keeping up to date with the use of technology is vital, it must be remembered that the design or revision of a production system is more than just a way to use technology, it encompasses the development of a socio-technical system that reflects the nature of people and the way they work together (see also Dilworth, 1992).

Numerous publications comparing the Japanese management techniques with western techniques suggest that management techniques are not transportable and that the success of techniques depends on the technological and cultural environments of each country. Other writers such as Schroeder (1993) and Creech (1994) cite several examples of where Japanese techniques have been successfully used in the United States. Our own view is that with some adjustments any technique can be applied in any country. After all the Japanese move to quality stemmed from work done by an American, Dr Deming.

Notwithstanding, a common theme on technology should include:

- Product technology development should be simultaneous (or concurrent) with process technology development during the innovation stage.
- There is a significant scope of managing process technology which

should be appropriate to individual requirements against the technological, social and cultural background.

The contents of the following section relate primarily to process technology.

Choice of technology

The choice of process technology is not governed by exact science but by a combination of a number of logical criteria. We can group these criteria into two categories such as:

General criteria
- Volume growth
- Variety growth
- Degree of sophistication

Local criteria
- Users' experience and skill
- Supplier partnership

The general criteria of choice is applicable to all companies and all countries. A chemical process such as a refinery or a spray drying tower benefits from economy of scale whereas discrete processes such as metal cutting or packaging can benefit from small capacity increments. A large machine like a high-speed packing line has the benefit of a lower capital cost per unit of capacity, but the choice of a high speed line must be balanced against the volume growth over the life cycle of the product, otherwise the machine will only be partly utilized during a single shift operation. Another general criterion of choice is the variety and the growth of a product. Smaller units of machines offer higher flexibility and lower unit manufacturing cost when the growth in product variation is high. Schonberger (1986) uses as an example the 'super machine cycle' and he clearly demonstrates the inflexibility and progressive inefficiency of increasing capacity in large increments.

The degree of sophistication of technology relates to one or more of the levels such as mechanization, automation and integration. With mechanization manual tasks are reduced but machine operators are required. Automation eliminates the operator for a small number of repetitive tasks (for example robotics). Integration is the higher level of automation with the combination of previously separated functions. A typical example of integration is a group of machines with reprogrammable controllers linked together by an automated materials-handling system and integrated through a central computer to enable the production of a variety of items. This is also known as flexible manufacturing systems or (FMS). A yet higher level of automated manufacture is computer-integrated manufacturing (CIM). This concept integrates information from product ideas to the output of high quality products. Information from the core manufacturing activities plus input where relevant from marketing, customer orders, suppliers, inward transportation, is used. CIM includes master scheduling, capacity planning, materials requirements, inventory control, purchasing, quality reporting and logistics. The higher the level of the sophistication of technology, the higher is the intensity of capital. High capital investment in technology (including high software cost) is justifiable for high-volume, high-life-cycle and high-added-value products (e.g. motor cars). Figure 6.6 illustrates the appropriate technology axis, along the diagonal of the product process matrix. The effectiveness of the choice of technology reduces with a gradual shift from the axis or diagonal.

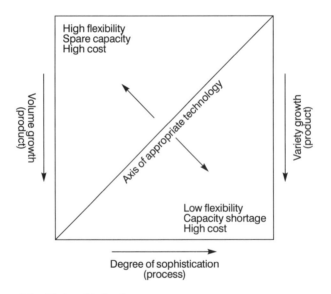

Figure 6.6 *Choice of technology*

The local criteria of choice depends on the cultural and technological environments of a country. The higher the technology the higher level of sophistication required of the workforce and the more efficient the infrastructure has to be (reliable energy sources and communications). The experience curve of the company on a known and proven technology should be a favourable criterion for choice. Where there is the necessary skill for managing technology then improved technology, as it becomes available, should be given serious consideration.

Partnership, with preferred vendors offer longer-term benefits and should influence the choice and confidence in technology. These benefits include:

- experience in proven technology
- standardization of equipment (especially attractive to multinational companies)
- involvement in the modification and design of equipment
- more rapid response from local representatives

The 'turnkey' approach practised in Japan, where one supplier is responsible for all the plant in a complete production line, also has its merits. This avoids one supplier blaming the other. However the supplier must have the expertise in all processes in the production line.

Evaluation of technology

The primary tool used by western companies in the evaluation of technology is the return on investment (ROI). ROI has received much criticism in recent years. Schonberger (1986) analysed the 'overstated role of capital' and argued that ROI should be revised to measure 'benefits of reduced variability'. Japan's investment in capital equipment in the last two decades has been twice that of its western competitors. Japanese investment in the training of workers (necessary to cope with higher-level technology) is also a lot higher per worker than in any other country. However it is their type of training that is important. Creech (1994) claims that in the United States for every graduate engineer there are eight lawyers, whereas in Japan for every lawyer there are 10 graduate engineers. It is axiomatic that lawyers do not add value to a process.

The newer technology (e.g. flexible manufacturing systems) could not show high returns and became difficult to handle financially in the western world. Another criticism is that investment decisions based on cost savings alone could lead to the continuation of historical constraints with 'out of date' equipment.

Hamel and Prahalad (1994) conclude that the corporate objective of ROI improvement (comprising two components: a numerator – net income – and a denominator – investment) has produced, in the USA and the UK, 'an entire generation of managers obsessed with denominators'. Denominator management is an accountant's short cut to asset productivity. They also warn that, in a world in which competitors are capable of achieving 5 per cent, 10 per cent or 15 per cent of real growth in revenues, aggressive denominator reduction under a flat revenue stream is simply a way to sell market share and the future of the company.

The rhetoric of the authors may be strong to make the point, but the message is clear. The primary goal of a company is the opportunity to compete in the future by creating new products and businesses rather than the need to meet short-term pay-back periods to make annual reports look good.

On the other hand, during the 1990s, it has to be admitted that manufacturing companies in the western world have faced a 'capacity glut' due to a combination of a number of factors:

- a recession which reduced consumer demand;
- aggressive restructuring and cost cutting initiatives which have identified redundant or poorly utilized plant;
- systematic re-engineering and continuous improvement programmes have increased the plant efficiency and thus releasing extra capacity;
- a relatively free movement of goods across countries which has encouraged out-sourcing and resulted in an imbalance of capacity.

Thus many companies are carrying high asset values amounting to between 25 and 40 per cent of the value of goods they produce and a significant amount of cash is locked in unutilized assets. Therefore there is a real need to provide high-level attention to improve the productivity of capital assets.

From the above analysis it is evident that evaluation of technology should be based upon not just ROI alone, but also a number of strategic criteria. Slack (1991) suggests a balanced approach by taking into account:

- the *feasibility* of investment – how difficult it is to install the technology;
- the *acceptability* of the investment – how much it gives competitiveness and gives a return on investment;
- the *vulnerability* of the investment – how much risk is involved in terms of what could go wrong.

Figure 6.7 shows a simplified diagram of the logical process for the selection and evaluation of appropriate technology. It is necessary to follow conventional capital investment appraisal procedures such as discounted cash flow (DCF) and net present value (NPV). The costs

MANUFACTURING FACILITIES   87

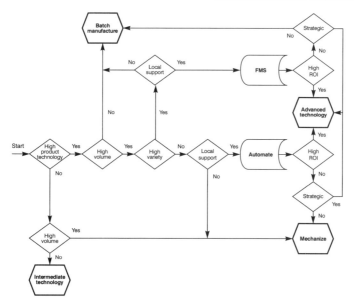

Figure 6.7   *Technology selection and evaluation*

must include 'hidden' elements such as commissioning, training, software development and support. Non-financial benefits should also be examined such as improved quality, saving of space, customer service, flexibility, avoidance of repetitive stress injury, lower stock holding and so on. If the return on investment (ROI) is less than the acceptable level of a conventional investment, then strategic considerations of the future (such as competitors' investment strategy) may be the decider when considering proposals for appropriate technology.

Foundation 12: Flexible manufacturing

Encouraged by the Japanese investment in flexibility and their success in the just-in-time strategy, flexibility has become one of the most fashionable of manufacturing virtues. However it also created confusion and ambiguity in the western world. There are instances of failures during the 1980s where companies invested in sophisticated FMS (flexible manufacturing systems) in pursuit of flexibility. At the other end of the scale all the attentions were given to organizational flexibility (e.g. cultural and skills integration between craftsmen and operators), producing limited success. It is important to clarify what we mean by flexibility in manufacturing, why we need flexibility and how we can improve flexibility.

What is flexibility?

Flexibility in manufacturing is the ability to respond quickly to the variations of manufacturing requirements in product volume, product variety and of the supply chain.

The variability in volume is demonstrated by product launching, seasonal demand, substitution and promotional activities. The changes in variety relate to increased number of SKUs (stock keeping units) in new products, distributors' own brands (DOB), etc.

The variations in the supply chain result from variability of lead times of both suppliers and customers, increased service level, change in order size, etc.

Slack (1991) distinguishes between range flexibility – how far the operation can be changed – and response flexibility – how fast the operation can be changed.

Why be flexible?

Flexibility is the ability to be responsive. In the past the manufacturers depended on a limited product range often supported by a protective market. Henry Ford and the days of the Model T car – 'You can choose any colour as long as it is black' – are long gone. Traditionally companies managed the variations in volume, variety and customer service by building stocks and/or excessive production capacity. This approach generated poor productivity and required heavy investment in stock piles of raw materials, work in progress and output stocks.

With the increased choices available to customers, often induced by competitors' marketing, the fragmentation of the market for increased product varieties will continue. Flexible manufacturing can offer a competitive advantage by giving customers greater choice of selection, and with less waiting time (Toyota's 72 hour car, see Chapter 2).

Within the factory flexibility offers the following benefits:

- Reduced departmentalization of equipment leads to reduced movement and handling of materials which should result in less direct labour.
- Reduced capital investment as fewer and less specialized machines are needed. Multifunctional machines should lead to greater machine utilization.
- Set-up or change over time should be reduced. One installation visited by us had reduced set-up time from eight hours to three minutes.
- Factory workers are required to be multiskilled and job enrichment is experienced. When workers have job satisfaction, productivity inevitably increases.

But there is little intrinsic merit in flexibility just for its own sake. Unlike cost, quality and service, companies do not sell flexibility. Flexibility within the plant is not of any concern to the customer. The customer does not care how a product is made as long it is received on time and to specification.

Flexibility is the shock absorber of manufacturing to provide continuous customer service under conditions of uncertainty and variety, see Figure 6.8.

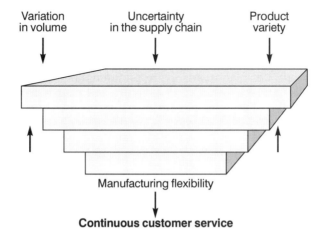

Figure 6.8 *Flexibility as a shock absorber*

How to improve flexibility

A change to a flexible system will require careful planning, capital investment and a change in attitude by workers. Intensive training of workers will be required. During the change-over productivity is likely to drop. Additionally, flexibility does imply some slack in the system. The first consideration then must be to be certain that there are benefits to be gained from changing from long runs to a flexible system. Additionally, flexibility does suggest the acceptance of some slack in the system. Even if flexibility is achieved without cost, it should not be wasted in areas where it could be avoided.

A flexibility strategy, if adopted, should be applied at different stages of the business process across all functions. It is not the responsibility of manufacturing alone and should span product design, process design, operations management, human resources, suppliers, systems flexibility and modular design.

Product design

One important factor affecting the decision to adopt flexibility or not is product design philosophy aimed at reducing complexity. The harmonization of products and materials can reduce the number of SKUs and production change-over times without reducing the 'variety' for customers. For example it is easy to standardize the shape and size of ice-cream cones and maintain different flavour and packaging so as to 'split' markets.

Process design

We discussed the rationale of FMS in the previous section. Although advanced 'flexible' plants with programmable controls (e.g. FMS, robotics, CAD/CAM) offer efficient product change-over, the high initial capital cost and inadequate local support may impede their implementation. However, simple modifications (such as quick change-over parts) can be easily built into the conventional production plant to improve flexibility.

Operations management

One much publicized approach of improving flexibility in current operations is SMED (single minute exchange of dies) developed for the Japanese automobile industry by Shigeo Shingo (1985). The SMED method involves the reduction of production change-over by extensive work study of the change-over process and identifying the 'in process' and 'out of process' activities and then systematically improving the planning, tooling and operations of the change-over process (see Figure 6.9). Shingo believes in looking for simple solutions rather than relying on technology. With due respect to the success of the SMED method, it is fair to point out that the basic principles are fundamentally the application of classical industrial engineering or work study.

Systems flexibility

The accuracy of planning data (e.g. stock records, production standards, capacity information), the reliability of products, plants and the planning tools (e.g. forecasting, materials requirements, production scheduling), worker training and attitudes, are all essential for improved systems flexibility for the total supply chain.

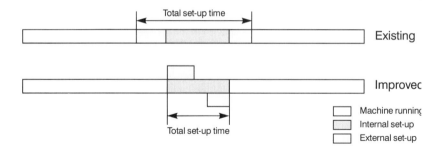

Figure 6.9 *Set-up time reduction*

Layout flexibility

The flexible design of facilities enables incremental changes in equipment or a layout to handle growth in production. Layout flexibility achieves two objectives – lower additional cost of facilities and minimum disruption of current operations. Figure 6.10 (from Tompkins, 1989) illustrates a flexible design of facility expansion. By locating all utility services (e.g. steam, air, water, power) and the main passage for material handling along the 'spine', the disruption to processing is minimized.

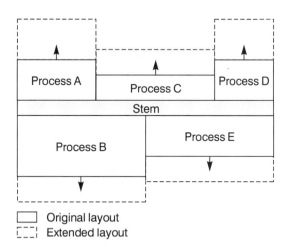

Figure 6.10 *Layout flexibility*

Foundation 13: Reliable manufacturing

Having established the selection and procurement of the appropriate manufacturing facilities for both the present requirements and longer-term competitive advantage, it is equally important to ensure dependability and reliability. If the plant is not reliable, in either availability or performance, companies must build up stocks or install excess capacity, thereby leading to poor capital utilization. Although the function of ensuring reliable manufacturing has been given different names to induce a higher profile (e.g. assets care, reliability engineering, etc.) it is best known as maintenance.

A maintenance strategy should cover three levels of action such as maintenance avoidance (longer term), maintenance reduction (medium term), and maintenance improvement (current or short term). Maintenance avoidance relates to longer-term 'right first time' measures at the selection and procurement stage of a capital project so that appropriate specifications and project management can ensure capital plant which is intrinsically reliable with associated low maintenance costs. Maintenance reduction relates to medium-term measures of existing equipment by continuous modifications of plant to eliminate weak points. Maintenance improvement relates to the policy, planning and control of the current maintenance activities.

Trends in maintenance

One traditional role of the maintenance function within a company up to the 1980s was viewed as a simple cost centre whose main contribution was to keep the factory plant running at a minimum cost. The trends away from the labour-intensive to the computer-controlled intensive production and from manufacturing for stock to just-in-time manufacturing have made efficient maintenance a key function. Maintenance is no longer a cost center, it is a competitive weapon for manufacturing.

The cost of maintenance should be perceived at two levels. The visible cost incurred by own labour, materials and third party maintenance is only the tip of an iceberg. The less tangible maintenance related costs (downtime, scrap, reworks, holding stocks of spare parts and even reserve machines, lost sales, poor workmanship, poor product quality and safety hazards) are many times more than the direct maintenance cost. Some of these costs can be measured, but some, such as lost sales due to poor quality or lack of reliable delivery, are unknown and unknowable.

Influenced by the change in technology, competition and work culture, the maintenance function has experienced a gradual transformation:

1960s: Breakdown maintenance
1970s: Time-based preventative maintenance
1980s: Predictive maintenance
1990s: Total productive maintenance and empowerment of work force

Breakdown (or unscheduled or reactive or unplanned corrective) maintenance is the repair of a piece of equipment or asset as the fault occurs. This is not always the cheapest option. Preventative maintenance, including lubrication and correcting settings, can prevent expensive repairs. Likewise minimum direct maintenance will result in reduced availability and reliability. It is not unknown for equipment to break down at the least convenient time!

Time-based preventative maintenance is carried out at regular fixed intervals or after a fixed cumulative output or cycles of operation and follows documented procedures. Lubrication is in this category and is essential for all mechanical machinery. Figure 6.11 shows a model that is often used to illustrate the optimum level of preventative maintenance.

Condition-based predictive maintenance requires the measurement and monitoring of key parameters to look for changes in characteristics which indicate that the equipment is approaching failure or not. Preventative action is then dependent on the analysis of the condition data. This approach is appropriate and effective for rotating equipment and critical plant. For example, vibration measurement (e.g. velocity in mm/second on rotating equipment which indicates faults in alignment, foundation or imbalance), see Figure 6.12.

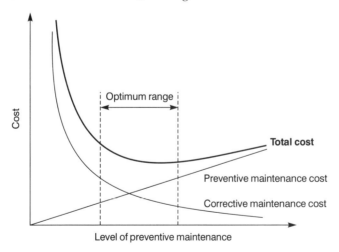

Figure 6.11 *Optimum maintenance*

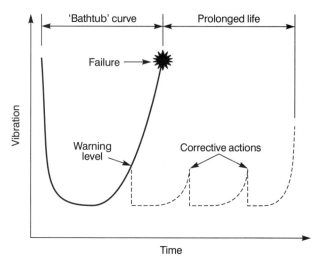

Figure 6.12 *Condition monitoring*

Inspection-based maintenance is aimed to establish the condition of the equipment by simple 'look listen and feel' inspection. This simple approach can be effective when administered by experienced engineers and craftsmen.

Total productive maintenance (TPM) is a proven Japanese approach to maximizing overall equipment effectiveness and utilization, and relies on attention to detail in all aspects of manufacturing. TPM includes the operators looking after their own maintenance and thus encourages the empowerment. TPM is described in more detail later.

Maintenance policy and infrastructure

The present experience both in Japan and the rest of the world has demonstrated the virtues of TPM. However it is also vital to establish the appropriate maintenance policy and infrastructure, both before and during the application of TPM principles, to sustain the reliability and safety of plant.

Although we have discussed the trend over the last four decades in maintenance policy, manufacturing companies are following (and quite rightly so) a mix of maintenance policies depending on the type and criticality of the equipment. There are elaborate tools and concepts such as RCM (reliability-centred maintenance) available to develop the optimum maintenance mix. Figure 6.13 shows a model to illustrate how a maintenance mix depends on the criticality of equipment.

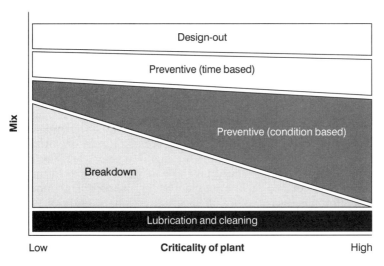

Figure 6.13 *The maintenance mix*

The criticality of a piece of equipment can be determined by the role of the equipment in the total process, repair expenditure incurred, and the production and quality consequences. The experience of the factory engineer and manufacturing manager should be sufficient to categorise each item of equipment and assess the criticality of that item. The following examples of maintenance mix could be used as a guide:

1. All equipment – lubrication and cleaning to be carried out to a formalized plan.
2. Critical equipment – assets with high production and quality consequences or repair expenditure to be put on condition-based predictive maintenance, e.g. rotary equipment should be on vibration monitoring.
3. Packing lines – to be on periodic inspection-based maintenance with proper checklists. Overhaul (part or full should be taken only on the basis of inspection observations).
4. Less critical equipment – assets with lower production consequences to be on formalized look, listen and feel inspection coupled with corrective action.
5. Services – same approach as production equipment but third-party contracts will play a more important role here. Statutory regulations (e.g. for pressure vessels) could determine time-based preventative maintenance.
6. Buildings – a long-term (e.g. 10 years) time-based preventative

plan for setting up a specified repair budget plan. Building maintenance should be contracted out.
7. Internal transport – a time-based preventative maintenance plan for this type of equipment (e.g. fork lift trucks) which should be carried out by specialist third parties.

In order to implement the maintenance mix, the company must provide maintenance infrastructure in the form of organization, workshops, engineering stores, planning procedures and information systems.

There are two aspects of maintenance organization – organization of skills and organization structure. Organization issues are discussed in Chapter 8. However, it is important to note that factories in the future will require fewer people but people with higher technical knowledge. It is therefore necessary to define basic educational standards, with the future in mind, for the recruitment of operators, team leaders and technicians to ensure that they will be capable of running and maintaining sophisticated machinery.

The maintenance workshop should contain machinery for emergency repairs, including welding equipment and instrumentation and monitoring equipment. Building maintenance, non-emergency repairs, and specialist maintenance should be undertaken by third parties wherever possible. Each site should have one engineering store with appropriate coding and layout for both general parts and machine spares.

The planning procedures should be designed to obtain permits (where required), checklists for inspection and lubrication, and scheduling of large repair work. Excessive planning, on the other hand, risks the danger of a slow response. Finally records should be kept of when each item was purchased and from whom, warranty details, and of what and when maintenance is carried out. Manufacturers' manuals should be filed and updated as required.

TPM (Total Productive Maintenance)

The use of the word 'maintenance' in the title is misleading. Total productive maintenance includes more than maintenance, it addresses all aspects of manufacturing. The two primary goals of TPM are to develop optimum conditions for the factory through a self-help people/machine system culture and to improve the overall quality of the workplace. It involves every employee in the factory. Implementation requires several years, and success relies on sustained management commitment. TPM is promoted throughout the world by the Japan Institute of Plant Maintenance (JIPM).

TPM is the manufacturing arm of total quality management (TQM) and is based upon five key principles:

1. The improvement of manufacturing efficiency by the elimination of six big losses.
2. The establishment of a system of autonomous maintenance by operators working in small groups.
3. An effective planned maintenance system by expert engineers.
4. A training system for increasing the skill and knowledge level of all permanent employees.
5. A system of maintenance prevention where engineers work closely with suppliers to specify and design equipment which requires less maintenance.

TPM requires the manufacturing team to improve asset utilization and manufacturing costs by the systematic study and the elimination of the major obstacles to efficiency. In TPM these are called the 'six big losses' and are attributed to (i) breakdown, (ii) set-up and adjustment, (iii) minor stoppages, (iv) reduced speed, (v) quality defects and (vi) start-up and shut-down.

The process of autonomous maintenance is to encourage operators to care for their equipment by performing daily checks, cleaning, lubrication, adjustments, size changes, simple repairs and the early detection of abnormalities. It is a step-by-step approach to bring the equipment at least to its original condition.

Some managers may hold the belief that in TPM 'you do not need experienced craftsmen or engineers and all maintenance is done by operators'. This is not true. The implementation of a maintenance policy with appropriate infrastructure is fundamental to planned maintenance. Planned maintenance is the foundation stone of TPM. However if the skill and education levels of operators are high then a good proportion of planned maintenance activities should be executed by operators after proper training. Cleaning, lubrication and minor adjustments together with an ability to recognize when a machine is not functioning correctly should be the minimum which is required of operators.

For TPM to succeed a structural training programme must be undertaken in parallel with the stages of TPM implementation. In addition 'one point lessons' can be used to fill in a specific knowledge gap. This uses a chart which is displayed at the workplace and describes a single piece of equipment and its setting or repair method.

Whilst great progress can be made in reducing breakdowns with autonomous maintenance and planned maintenance, 'zero breakdowns' can only be achieved by the specification of parts and equipment which are designed to give full functionality and not to fail. All engineers and designers of the user company should work concurrently with the suppliers of equipment to achieve a system of maintenance prevention.

Although there is a special emphasis of input by different employees to different aspects of TPM (e.g. 'six big losses' for middle management, 'autonomous maintenance' for operators, 'planned maintenance' for middle management, 'maintenance prevention' for senior management) TPM involves all employees and the total involvement is ensured by establishing TPM work groups or committees. Figure 6.14 illustrates an example of a TPM organization.

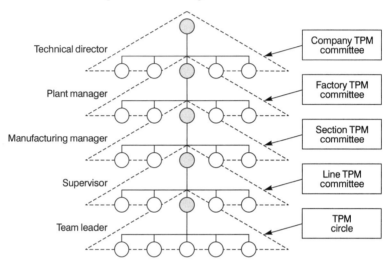

Figure 6.14 *TPM organization*

To summarize, TPM is a factory-wide continuous improvement programme with particular emphasis on changing the culture of the shop floor through improved attitudes and skills. TPM progress is measured by the stages of autonomous maintenance completed, and visible progress is also seen in the higher reliability of equipment, reduction of waste and improvements in safety statistics.

Foundation 14: Manufacturing performance

There may be a minority group of 'behaviourists' who consider the measurement of performance is an impediment to the creativity and empowerment of employees. However in the real business world, including Japan, successful companies are pursuing competitive parameters of manufacturing by continuous measurement, monitoring and improvement programmes. A recent case example, Adler (1993) has clearly demonstrated that by defining performance standards, quality and productivity at the Fremont plant (a GM-Toyota joint venture) went from worst in the General Motors' quality league (before Toyota got involved) to best shortly after Toyota became involved. The Toyota approach was centred on teams, with each team responsible for setting measurable objectives in the areas of quality, cost, production and safety (see also Creech, 1994). It is now well documented that a faster rate of progress has been successfully sustained in companies where work groups have been involved in setting, monitoring and improving their own work standards.

Having established the strategies of sourcing, appropriate technology, flexibility and reliability it is imperative to measure the performance of manufacturing facilities in order to monitor how the strategies are delivering the expected competitive advantage. If it cannot be measured there is no method of determining if an improvement has taken place. Or as F.W. Taylor once said, 'if it can't be measured it can't be managed'. There are several benefits arising from manufacturing performance efforts.

First, by defining and measuring the key parameters, the company can identify areas for improvement and take early corrective action. Second, by comparing performance across units, organizations and industries, management can make decisions on both operational improvement and research needs. Any benchmarking exercise (whether 'internal' or 'external') cannot be effective without measured performance. Third, a planning and scheduling scheme is only as reliable as the data used and thus only measured performance data can ensure a dependability in customer service. Fourth, performance measurement when carried out by teams made up of all levels of employees unites labour and management in seeking real productivity gains. Fifth, measuring performance also gives an organization the evidence it needs to quantify and celebrate gain, whether it is for employee incentive schemes or to demonstrate gains to interested stakeholders.

Our model of manufacturing performance has two elements – manufacturing efficiency and manufacturing effectiveness. Manufacturing efficiency is internal, factory focused and relates to the utilization of resources (namely assets, materials, energy and people). Manufacturing effectiveness is external, customer focused and relates to quality, delivery and environment. Slack (1991) gives parameters of manufacturing advantage as:

- Doing it right – the quality advantage
- Doing things fast – the speed advantage
- Doing things on time – the dependability advantage
- Changing what you do – the flexibility advantage
- Doing things cheap (without sacrificing quality) – the cost advantage

Although they are not strictly identical, manufacturing efficiency is responsible for the 'cost advantage' and manufacturing effectiveness is responsible for the other 'advantages'.

Manufacturing efficiency

The measurement, monitoring and improvement of manufacturing efficiency are strongly established functions within the field of traditional industrial engineering. However, many companies – perhaps influenced by the tide of fashionable TLAs (three-letter acronyms) abandoned the professionalism of 'efficiency experts'. Arguably, some companies misinterpreted W.E. Deming's statement, 'Eliminate numerical quotas for the work force: as they disregard quality and put a ceiling on production'. After all Deming also said, 'Use statistical methods to find the trouble spots'. There is some history of the abuse of labour standards to determine incentive payments (time studies and Taylorism). The leverage of efficiency has certainly moved away from labour with increased mechanisation and automation of repetitive work, but it has shifted to the efficiency of assets, materials and energy. Measurement is still important, the issue now is to measure what matters.

Labour productivity

Labour productivity can be measured and expressed in one of three ways.

1. Standard hours, where standard actual hours and output for a task or product are compared to actual time taken.
2. Input of resource, such as employees per unit or employee hours per tonne.
3. Output such as units of output per employee or tonnes per employee hour.

The need and choice of work measurement tools (such as time study, pre-determined motion time systems, activity sampling, etc.) will depend on the nature of operations and the importance of labour in the total performance. Labour standards are valuable data for resource planning, standard costing and activity-based costing. Labour is also the most controllable of all manufacturing resources.

Materials productivity

In terms of cost the most important resource is materials. In an FMCG business raw and packaging materials account for 70 to 80 per cent of the ex-works cost of a product. Because of the large number of line items or SKUs (stock keeping units) involved in the bill of materials of a product, a popular method of expressing materials productivity has been financial indices such as the ratio of standard material cost to actual material cost. Cost performance provides an indicator of the trend of materials cost for the product by using a common unit for all materials (e.g. the cost), but it fails to give the yield or productivity of a specific material.

Other measures of material productivity are materials yield and loss rate by each material where:

$$\text{Material yield} = \frac{\text{Theoretical consumption}}{\text{Actual consumption}} \times 100$$

$$\text{Loss ratio} = \frac{\text{Reject or losses}}{\text{Actual consumption}} \times 100$$

The data collection and measurement of materials productivity can be complex and this can lead to inaccuracies. Hence companies achieve effective results by restricting the monitoring of materials productivity to key items selected by Pareto analysis (the 80/20 rule which says that 80 per cent of the problems can be attributed to 20 per cent of the causes). The reliable information on materials costs and productivity not only assists continuous improvement programmes but also leads to longer-term decisions on purchasing of materials and design/formulations of products.

Energy efficiency

Energy management benefits in two areas – cost reduction (energy accounts for 15 to 30 per cent of the conversion cost of a product) and environmental protection. There are a number of factors which would influence a longer-term improvement (e.g. choice of fuel, type of equipment, generation process, etc.). However, by eliminating losses and monitoring some simple indices, companies can implement a continuous energy efficiency programme. The main indices for an FMCG plant are:

- **Steam:** Tonnes of steam per tonne of product
- **Electricity:** kWh (kilo watt hour) per tonne of product
- **Fuel:** kg of fuel per tonne of product
- **Overall:** GJ (gigajoule) per tonne of product

102 TOTAL MANUFACTURING SOLUTIONS

It is important that the above indices are used for monitoring the trend of the factory. Only and if inter-factory benchmarking is considered then appropriate metering of utilities would be necessary.

Plant efficiency

As the manufacturing operations have progressively become more process and equipment dependent, the utilization and efficiency of plant have become the most important driving element of manufacturing performance. A properly designed and administered plant efficiency scheme offers broad ranging benefits and a comprehensive manufacturing performance system:

- it provides information for improving asset utilization and thus reduces capital and depreciation costs in the longer term;
- it highlights equipment faults and thus improves plant reliability and contributes to the designing out of weak points;
- higher plant efficiency improves labour productivity by producing higher output without increasing the number of employees required;
- plant efficiency information focuses on downtimes caused by services (steam, air, water, power) and a higher plant efficiency results in a higher energy efficiency;
- it identifies the direct losses caused by poor quality of materials and thus improves material productivity and product quality;
- it provides essential and reliable information for capacity planning;
- it provides information on downtime due to shortage of materials and thus improves materials planning;
- it provides information for effective scheduling of plant to shorten lead times, e.g. by changing the operation from two to three shifts.

There are many ways to calculate plant efficiency (see Shirose, 1992 and Hartmann, 1991). The success factor of such a plant efficiency and capacity evaluation scheme (PEACE) is that the parameters must be well defined and measurements should not rely heavily on recorded downtime. Figure 6.15 shows a basic model of plant time analysis of a simple but robust scheme with the following definitions:

Maximum time (M) defines the total plant time in the reporting period, i.e. 24 hours per day = 168 hours per week.

Planning time (P) is the realistic time during which the plant can be planned for production, and is established each year for annual and longer-term planning, after taking out unplanned time (due to holidays and local restrictions) from maximum time.

Operating time (O) is the time during which the plant is actually

MANUFACTURING FACILITIES   103

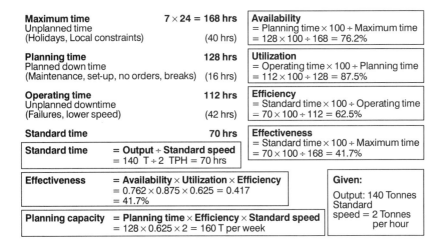

Figure 6.15 *Plant efficiency and capacity evaluation (PEACE)*

scheduled and manned to operate for production purposes. Planned downtime (due to no order, planned maintenance, set up and breaks) constitutes the difference between planning time and operating time.

Standard time (S) is calculated from good output and standard speed as:

$$\text{Standard time} = \text{Good output} / \text{Standard speed}$$

where standard speed is the optimum speed of a machine for a particular product without any loss of efficiency.

Unplanned downtime is the difference between operating time and standard time and comprises failures, lower speed, quality losses and other unrecorded losses.

Four indices are suggested for reporting both plant performance:

$$\text{Availability} = \frac{P}{M} = \frac{\text{Planning time}}{\text{Maximum time}}$$

Availability indicates what proportion of maximum time can be made available for annual and long-term capacity planning.

$$\text{Utilization} = \frac{O}{P} = \frac{\text{Operating time}}{\text{Planning time}}$$

Utilization indicates what proportion of planning time has been utilized for actual productive operations.

$$\text{Efficiency} = \frac{S}{O} = \frac{\text{Standard time}}{\text{Operating time}}$$

Efficiency demonstrates how efficiently the output has been delivered during the operating time by minimizing avoidable losses.

$$\text{Effectiveness} = \frac{S}{M} = \frac{\text{Standard time}}{\text{Maximum time}}$$

$$= \text{Availability X Utilization X Efficiency}$$

Effectiveness is the true measure of good output that can be obtained from plant in the maximum time available. This is the net result of all measures. The usual effectiveness range of a plant is between 40 to 60 per cent

Planning capacity measures the output that could be achieved by a plant when operated at a given efficiency during planning time, for example:

Planning capacity = Standard speed x Given efficiency x Planning time

Manufacturing effectiveness

Internal

Manufacturing effectiveness is primarily external and customer focused. However there are some useful parameters (see Table 6.2) and indices of manufacturing effectiveness which are internal and factory focused. Table 6.3 shows the key external parameters and indices for manufacturing performance that customers are looking for. We have covered some of these in previous chapters.

Table 6.2 *Manufacturing effectiveness – internal*

Parameter	Indices
Planning	Actual output versus planned output (by product)
Inventory/stock	Stock cover or stock turn Stock value as percentage of total sales

	Safety	Loss time accident rate percentage First aid attendance
	Environment	Percentage compliance of emission rate
	Product Quality	Defined waste as percentage of total sales
	Training	Training days per employee Training cost as percentage of total sales

Table 6.3 *Manufacturing effectiveness – external*

Parameter	Indices
Customer service	Order fill percentage Lead time (days) On time delivery percentage
Quality	Goods returned as percentage of total sales Complaints as percentage of number of invoices
Innovation	New product sales as percentage of total sales

Performance improvement

There is a risk of neglecting the opportunities of incremental improvement of the performance of the existing systems by down-grading the approach as 'tinkering'. While the vision must be to look for step changes in competitive advantages by strategic initiatives, the opportunities for continuous improvement cannot be ignored.

The whole theme and emphasis of this book is to identify the right balance of opportunities by a self-analysis in order to see the wood for the trees.

A continuous improvement programme can be in three stages.

1. Monitoring – the continuous monitoring of appropriate performance indices, ideally by professional industrial engineers, and regular discussions and corrective actions by all functions.
2. Special studies – when the trend of performance indicators shows

chronic problems, special studies (such as video based machine motion analysis, and failure mode effect analysis) are carried out by a team or task force.
3. Change programmes – in order to sustain the improvements in manufacturing performance, a company-wide change programme (such as total quality management or total productive maintenance) is implemented.

Summary

As Aristotle said, 'Give me a lever long enough and I will move the world'. This chapter covers the five foundation stones of the 'core' pillar of manufacturing. This is a big chapter with much detail.

Management of manufacturing facilities is the forgotten link in corporate competitive strategy, yet in manufacturing companies it is the factories that employ the major share of the companies' assets and the majority of the people.

To gain, and to maintain, a competitive advantage it is essential to know and to understand each of the manufacturing facilities foundation stones.

Foundation 10

Sourcing strategy deals with where we get our materials and people, and whether we make or buy. Sourcing requires a formal strategic plan. This plan must be in tune with the overall corporate strategy. We recommend a project team approach to develop a formal strategic sourcing plan.

Foundation 11

In this section we discuss the pros and cons of technology. Without proper planning – including product, process, choice of technology and evaluation of options – technology can well become a costly millstone. There are many well publicized 'green-field' hi-tech factories which have proved to be white elephants. In this section we show how to plan so as to avoid costly mistakes which are hard to undo.

Foundation 12

Here we examine the fashionable subject of flexible manufacturing. We warn that there is little merit in flexibility for its own sake. After all the customer does not care how the plant is organized providing the product is to specification and received on time. We point out that the crucial issues with flexibility are the training and attitude of workers. The workers have to be trained, and receptive, or flexibility will result in confusion and a drop in productivity.

Foundation 13

For this foundation stone we examine the various levels and ways in which maintenance can be planned so that the manufacturing process becomes more reliable. The highest level of maintenance is total productive maintenance (TPM). We describe TPM in some detail as we firmly believe that maintenance is a major issue in giving a company a competitive edge. The success of TPM is dependent on the quality and attitude of the workers.

Foundation 14

In this section on manufacturing performance we revisit F.W. Taylor's statement, 'If it can't be measured, it can't be managed'. If the system does not have standards which are quantifiable how do we know if there has been an improvement, and how do we know if we are not actually losing ground? This section gives some key measures. Again we say unless we use the measurements there is no point in taking them.

In all our activities we must be conscious of the aim. The aim is to add value and to reduce costs.

References

Adler, P.S. (1993) Time and Motion Regained. *Harvard Business Review*, Jan-Feb, 76, No. 1.
Brown, R.M. and Blevins, T.F. (1989) Should America Embrace Japanese Management Techniques? *Advanced Management Journal*, Winter, 22-31.
Creech, B. (1994) *The Five Pillars of TQM*. Truman Talley.
Dilworth, J.B. (1992) *Operations Management: Design, Planning and Control for Manufacturing and Services*. McGraw-Hill.
Hamel, G. and Prahalad, E.K. (1994) Competing for the Future. *Harvard Business Review*, Jul-Aug, 72, No. 4.
Hartmann, E. (1991) How to Install TPM in your Plant. 8th International Maintenance Conference, Dallas, 12-14 November.
Hill, T. (1985) *Manufacturing Strategy*. Macmillan.
Miller, J.G., De Meyer, A. and Nakana, J. (1992) *Factories of the Future*. Irwin.
Schonberger, R. (1986) *World Class Manufacturing*. Free Press.
Schroeder, R.G. (1993) *Operations Management's Decision Making in the Operations Function*. McGraw-Hill.

Shingo, S. (1985) *A Revolution in Manufacturing*, The SMED System. Productivity Press.
Shirose, K. (1992) *TPM for Workshop Leaders*. (Trans. B. Talbot) Productivity Press.
Skinner, W. (1969) Manufacturing is the Missing Link of Corporate Strategy. *Harvard Business Review*, May-June, 136-145.
Slack, N. (1991) *Manufacturing Advantage*. Mercury.
Stalk, G. Jr (1988) Time – The Next Source of Competitive Advantage. *Harvard Business Review*, Jul-Aug, 66, No. 4.
Tompkins, J.A. (1989) *Winning Manufacturing*. Engineering and Management Press.
Wheelwright, S.C. and Hayes, R.H. (1985) Competing Through Manufacturing. *Harvard Business Review*, Jan-Feb, 63, No. 1.
Wild, R. (1995) *Production and Operations Management*. Cassell.
Yip, G.S. (1992) *Total Global Strategy: Managing for Competitive Advantage*. Prentice Hall.

7
Procedures

The golden rule is that there are no golden rules.
George Bernard Shaw

Chapter 6 discussed manufacturing facilities, the 'hardware' of manufacturing. We now turn to the 'software', that is systems and procedures. This chapter considers the three systems and procedure foundation stones:

15. Quality management
16. Financial management
17. Information technology

Earlier we considered the implications of issues such as shorter innovation cycles, more stringent product specifications, asset performance, standard on-time delivery to customers, closer relations with both suppliers and customers, and so on. Attention to these issues accentuates the need for new integrated and flexible management control systems.

Another important issue is improving the financial performance of the company. Under pressure to participate in fashionable improvement activities, or to become involved with the newest business wisdom, management may lose sight of the real issue – improving profitability. The whole aim of this book is to improve profitability of an organization by concentrating on adding value through manufacturing endeavours.

Foundation 15: Quality management

What is quality?

Quality has two levels, a basic level and a higher level. At the basic level common definitions 'fitness for purpose', 'getting it right first time', and 'right thing, right place, right time' apply. (These definitions have all been so over used that they are almost clichés.) An understanding of what we mean by basic level and higher levels of quality can best be explained by illustration.

Consider a bus service. What as passengers are our basic requirements? First, unless the bus is going more or less where we want to go, we won't catch it. The second requirement is timing – usually we have a time frame by which we judge a bus service. If we start work at 9 a.m. unless the bus gets us to the office before 9 we won't catch it. Another consideration will be cost. Therefore the basic requirements in this example would be the route, the time and the cost, and depending on alternatives we would probably rank them in that order.

A bus service could meet all these requirements, (right thing, right place, right time, and right cost), but still not be a quality service. If the service was unreliable, (sometimes late, sometimes early, sometimes did not keep to the route) then we would not consider it a reliable service. But supposing the bus met all our basic requirements, got us to work on time every time and at a reasonable cost, *but* it was dirty, the driver was surly, the seats were hard and it leaked exhaust fumes. Then although it met our basic requirements there is no way we would describe it as a quality service.

In other words to meet our perception of quality there are certain basic requirements that have to be met, and there are certain higher order requirements that have to be met. In this case we would expect polite service, a clean bus, reasonably comfortable seating and certainly no exhaust fumes. A truly high quality service would mean that the bus was spotlessly clean, had carpet on the floor, and had piped music as well as all the other attributes. But no matter how comfortable the ride, how cheap the fare, unless the bus is going our way we shan't be interested in catching it.

To have your product described as a quality product, the customer will expect higher level benefits. These higher level benefits are what gives an organization a competitive edge, and often the difference costs very little to achieve.

Hierarchy of quality

With the subject of quality, like many management subjects such as marketing, and strategic management, a number of technical terms have evolved. In some cases rather than helping us to understand the underlying concepts or techniques, technical terms tend to add a further complication to our understanding. Often also the terms used are given different connotations by different people, the meanings become blurred, and terms become interchangeable. In this section we discuss the various ways in which quality can be managed. We also discuss the strengths and weaknesses of each method. For these reasons we have developed a hierarchy of methods of quality management. Our hierarchy approximates the evolution of quality management from simple testing to a full total quality management system.

Quality by inspection

Traditionally in manufacturing the concept of quality was conformance to certain dimensions and specifications, the cliché being 'fitness for purpose'. Quality control was achieved by inspection and supervision. This, the most basic approach to quality, can be labelled as quality by inspection.

Quality by inspection, if every deviation from standard is detected by the inspector before despatch, will at least provide the customer with an acceptable product. Although an acceptable product might satisfy the customer it is not likely to encourage customer loyalty. It is our contention that a competitive edge can be only be gained by providing the customer with more than they expect.

Quality inspection is an expensive method of achieving a basic level of quality. It requires the employment of people to check on the operators. Inspection and supervision do not add value to a product, they merely add to the cost!

The stage of production where the inspection takes place is important. If the only inspection is at the end of the production line then, if deviations from the standard are discovered, the cost of reworking could well double the cost of the item. If a deviation from standard is not detected, the final inspector becomes the customer, by which time it is too late. If the product is found to be below standard by the customer, the manufacturer has the problem of putting it right. Putting right could include the cost of scrapping the unit and giving the client a new one, or in extreme cases a total product recall with all the costs and loss of consumer confidence that this entails.

Quality inspection at a more advanced level includes checking and testing at various stages of production so that errors can be detected early and remedial action taken before the next stage of the process takes place. At a still higher level of inspection materials are inspected on receipt and then probably tested again before being drawn from the store. Of course all these tests and checks take time and cost money. The cost is easy to quantify when the checks are carried out by people whose prime job is to test and check the work of others.

It is our contention that when people know everything they do is subject to testing and checking, then the onus is no longer on them to get the job right first time and they come to rely on the inspector. We believe that the inspector or supervisor will be conditioned to find a percentage of errors, after all that is the main reason for employing inspectors. This attitude will be reinforced further by an error percentage being built into the standard costs. Thus, a level of error becomes accepted and is built into the cost of the product.

The costs of relying on inspection by people other than the operator are therefore two-fold:

1. A level of error becomes accepted as standard and is included in the price, and
2. Inspectors do not add value to the product. Inspectors are an added cost.

The next stage above quality inspection can be designated quality control.

Quality control

With quality control, the aim is not only to monitor the quality at various stages of the process but to identify and eliminate causes of unsatisfactory quality so that they don't happen again. Whereas inspection is an 'after the fact' approach, quality control is aimed at preventing mistakes. With quality control, you would expect to find in place drawings, raw material testing, intermediate process testing, some self-inspection by workers, keeping of records of failure, and some feedback to supervisors and operators of errors and percentage of errors. The end aims are to reduce waste by eliminating errors and to make sure that the production reaches a specified level of quality before shipment to the customer.

Quality assurance

Quality assurance includes all the steps taken under quality control and quality inspection. It includes, where appropriate, the setting of standards with documentation for dimensions, tolerances, machine settings, raw material grades, operating temperatures and any other safety quality or standard that might be desirable. Quality assurance would also include the documentation of the *method* of checking against the specified standards. Quality assurance generally includes a third party approval from a recognized authority such as the ISO. However ISO accreditation in itself does not suggest that a high level of quality has been reached. The only assurance which ISO accreditation gives is that the organization does have a defined level of quality and a defined procedure which is consistently being met. With quality assurance one would expect to move from detection of errors to correction of process so as to prevent errors. One would also expect a comprehensive quality manual, recording of failures to achieve quality standards and costs, use of statistical process control (SPC), and the audit of quality systems.

Total quality management

The fourth and highest level in our hierarchy of quality is total quality management. The lower levels – quality inspection, quality control and quality assurance – are aimed at achieving an agreed consistent level of quality, first by testing and inspection, then by rigid conformance to standards and procedures, and finally by efforts to eliminate causes of errors so that the defined accepted level of quality will be achieved. This is a cold and sterile approach to quality. It implies that once a sufficient level of quality has been achieved, then apart from maintaining that level which in itself might be hard work, little more need to be done. This is often the western approach to quality and has its roots in Taylorism (see Taylor, 1947). Taylor believed in finding the 'best method' by scientific means and then establishing this method as the standard. This approach is top down, the bosses determine the level of quality to be achieved, and then the bosses decide on the best method to achieve the desired level of quality. Control methods of inspection and supervision are then set in place to ensure that the required level of quality is maintained. This does not mean that management is not taking into account what the customer wants or is ignoring what the competition is doing. It just means that they, as managers, believe they know what is best and how this can be achieved. To this end, supervision and inspection become an important method of achieving the aim with little input expected from the workers.

Total quality management is on a different plane. Total quality management does, of course, include all the previous levels of setting standards and the means of measuring conformance to standards. In doing this, SPC will be used, systems will be documented, and accurate and timely feedback of results will be given. With TQM, ISO accreditation might be sought, but an organization that truly has embraced TQM will not need the ISO stamp of approval.

Any organization aspiring to TQM will have a vision of quality which goes far beyond mere conformity with a standard. TQM requires a culture whereby every member of the organization believes that not one day should go by without the organization in some way improving the quality of its goods and services. The vision of TQM must begin with the chief executive. If the chief executive has a passion for quality and continuous improvement, and if this passion can be transmitted down through the organization, then, paradoxically, the ongoing driving force will be from the bottom up.

Generally it is the lower-paid members of the organization who will physically make the product or provide the service, and it is the sum of the efforts that each individual puts into their part of the finished product which will determine the overall quality of the finished article. Likewise, generally it is the lower-paid staff members, such as shop assistants, telephone operators, and van drivers who are the contact point with the customer, and the wider public. They, too, have a huge part to play in how the customer perceives an organization. It is on the lower level that an organization must rely for the continuing daily level of quality. Quality, once the culture of quality has become ingrained, will be driven from bottom up, rather than achieved by direction or control from the top. Management will naturally have to continue to be responsible for planning and for providing the resources to enable the workers to do the job. But, unless the factory operators, the telephone operators, the cleaning staff, the sales assistants, the junior accounts clerk, and the van driver are fully committed to quality, TQM will never happen.

TQM, however, goes beyond the staff of the organization – it goes outside the organization and involves suppliers, customers and the general public.

Once a relationship has been built with a supplier, that supplier is no longer treated with suspicion, or in some cases almost as an adversary. Instead of trying to get the best deal possible out of the supplier, the supplier becomes a member of the team. The supplier becomes involved in the day-to-day problems and concerns of the organization and is expected to assist, help and advise. The supplier becomes part of the planning team. Price and discounts will no longer be the crucial issues, delivery of the correct materials at the right time will be the real issues, and suppliers will be judged accordingly. Once a supplier proves reliable, the checking and testing of inwards goods will become less crucial. Ideally, the level of trust will be such that the raw materials can be delivered direct to the operator's work place rather than to a central store.

Consider the difference to your organization if the raw materials were always there on time, were of the right quantity and quality, and were delivered to the operator's work place and not to a store; each operator knew the standards and got the job right first time every time; and so on right down the line. Then the organization would not need anyone involved in checking anyone else's work. Supervisors and middle management would no longer be policing each step of a job.

At the end of the process is the customer. TQM organizations are very customer-conscious. As the supplier is regarded as part of the team so too, is the customer. This is more than just wishy-washy slogans such as 'the customer is always right'. This means really getting alongside the customer and finding out exactly what they want. The ultimate is that the customer, like the supplier, becomes part of the process.

An example of the way the world is moving can be found with Toyota where the aim is the 72 hour car which we described in Chapter 2. To recap, with the 72 hour car the customer orders a new vehicle, the materials are ordered and the car is made and delivered to the customer, all within 72 hours. This allows the customer, within a range of options, to select the car of their choice, and the customer really does become part of the supply chain. The customer's order goes direct on-line to the suppliers and to the factory. Thus the customer triggers the raw material order for all the components required for the car and also the customer's order updates the manufacturing schedule for the factory. Taiichi Ohno of Toyota says that his current project is 'Looking at the time line from the moment the customer gives us an order to the point where we receive the cash. And we are reducing the time line by removing the non-value wastes.'

What does this mean? It means no more raw material stockpiling, no more stocks of finished goods, reduction in needs for capital, storage space, and insurance, and it means that the customer is getting what she or he really wants (such as colour, upholstery, sound system, engine size, and countless other options as specified by the customer). Obviously, a system such as the Toyota process does not, and cannot, make allowances for mistakes. A system such as this relies on good planning by management, quality designed into the product, well-trained workers who are empowered to work as a team, suppliers who are trusted to supply when required and who are also part of the team, an integrated computer system, and, as Taiichi Ohno says, the elimination of non-value wastes.

We are now then looking at a totally new type of organization: the old bureaucratic style of management, with the associated rules relating to span of control, appraisal systems, and incentive schemes is simply no longer appropriate. Instead, organizations have to be designed around the process. For example, instead of having a centralized purchasing department, why could not the operator, or a group of operators on the shop floor, phone or fax through the daily order to the supplier (and for the materials to be delivered directly to the line rather than to the store). If each group of operators around a process were working as a team, why would a large central human resources department be needed? Certainly, the operating team itself would not need a supervisor. Maybe a team leader would be necessary to hurry management along and to ensure that management planning was sensible. The aim here is not for the front-line operators to be working harder but for them to take control and accept responsibility for their operation. It does not mean fewer people turning out more, but it does mean the elimination of several levels of management and it does get rid of the matrix of responsibility for human resource and other 'service' or staff departments as shown on the old-fashioned organization charts. With fewer levels of management, communication becomes less confused, and responsibilities (and areas of mistakes) become much more obvious.

For TQM to work, a company has to go through a total revolution. Many people, especially middle managers, have to be won over. Workers, too, have to want to accept responsibility. TQM will mean a change of culture.

Figure 7.1 *Steps of TQM*

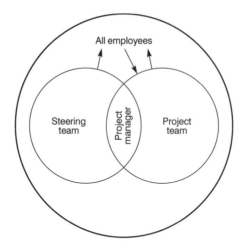

Figure 7.2 *TQM project organization*

The cost of TQM can be measured in money terms. The emphasis will be on prevention rather than detection, thus the cost of supervision and inspection will go down. Prevention cost will go up because of the training and action-orientated efforts. But the real benefits will be gained by a significant reduction in failures – both internal (e.g. scrap, rework, downtime) and external (handling of complaints, servicing costs, loss of goodwill). The total operating cost will reduce over time (say three to five years) as shown in Figure 7.3.

Problem-solving tools

It is important to apply common problem-solving tools and techniques across the organization, not only for training but also for effective com-

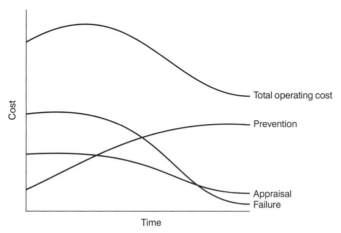

Figure 7.3 *Cost of TQM*

munications. Oakland (1992) proposes a number of problem solving tools, including the six tools of Bill Conway. In his capacity as the President of Nashua Corporation, USA, Conway demonstrated the six tools for quality improvement:

- Human relations skills
- Statistical surveys
- Simple statistical techniques
- Statistical process control (SPC)
- Imagineering
- Industrial engineering

Of the above tools, probably the most frequently applied is SPC and most neglected is industrial engineering. It should be evident from Figure 7.4 that industrial engineers usually apply most of the SPC tools in methods study and thus can play an important role in the TQM programme.

Implementation of TQM will require total and highly visible commitment by senior management, clear communication and an ongoing edu-

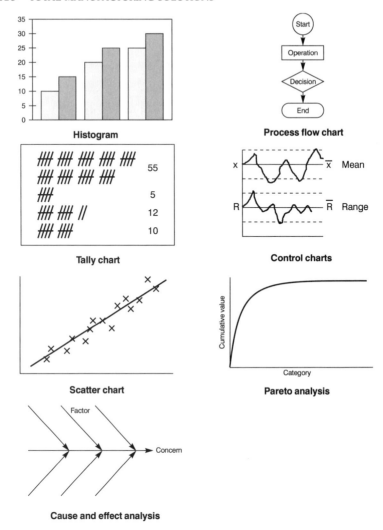

Figure 7.4 *The SPC tools*

cation programme. Everyone in the organization must go through a formal training process. To get the TQM wagon rolling will need considerable effort and dedication. This is probably best achieved by forming a project team made up of representatives from each function, or department, of the organization. Ideally project members will be volunteers. The key members of the project team will need specific training, not only in quality techniques, but in change management.

Following the formal launching of TQM the implementation process becomes a continuous improvement programme. It is not well understood in many organizations that TQM is a journey, not a destination! As shown in Figure 7.5, the improvement curl is a continuous loop cycle of plan, do, check and action. Only an effective co-ordination of these four components will result in sustainable competitive advantage of quality.

To maintain a wave of interest in the programme it may be necessary to dedicate the effort to the pursuit of an approved accreditation such as ISO 9000 or an award such as the Baldrige (in the USA) and derivatives of the Baldrige award in other countries. ISO 9000 accreditation, or achievement of an award such as a Baldrige, does not signify the end of

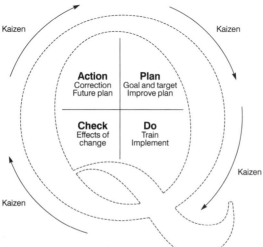

Figure 7.5 *TQM improvement circle*

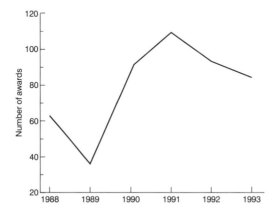

Figure 7.6 *TQM burnout?*

a journey, only a milestone. There is some indication that companies are moving away from TQM, perhaps they are viewing it as an internally focused initiative, or in some cases they might see TQM as just another management technique that didn't work (see Figure 7.6). In companies that lack the commitment to strive for TQM then ISO 9000 might seem to be a easier option. Then too it is likely that customers, especially export customers, might request the ISO 9000 stamp of approval.

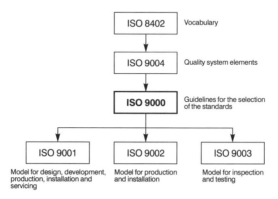

Figure 7.7 *ISO 9000 series*

ISO 9000 and total quality management

Total quality management means more than just the basics as outlined in BS 5750 or ISO 9000. In some ways, BS 5750/ISO 9000 could be seen as running contrary to the philosophy of TQM.

It is important to recognize the limitations of the ISO 9000 series. They are not and do not profess to be a cure for a business's ills. Many companies have misguidedly expected that by adopting an ISO 9000 standard they will achieve success comparable to that of the Japanese:

> One must not forget that the ISO 9000 standards did not exist when the Japanese quality performance improved so spectacularly: many Japanese firms did not need such written standards, and probably still do not.
>
> Sayle (1991)

BS 5750 standards had their origins with the UK defence standards which were introduced in the 1960s to improve the quality of materials supplied to the armed forces and to reduce the incidence of failure in the field. The British Ministry of Defence set standards which suppliers had to demonstrate they could meet before they were awarded a contract.

Once a supplier had gained a contract, they were subject to government audit. All in all it was a most expensive way to make sure that the army, navy and air force were getting what they paid for. During the 1980s, the UK defence standards were replaced by NATO equivalents.

Industry realized that there were benefits in the approaches used by defence, and British Standard BS 5750 was introduced in 1979. The International Standards Organization 9000 series is the most recent international standard. These standards are based on BS 5750, and ISO 9000 is today accepted as the world standard. To gain certification, an organization has to meet rigorous standards and satisfy a third party, the accreditation authority.

ISO, although it was originally intended for manufacturing industries, is becoming increasingly used by service industries, although some of the terms used by ISO need some translation for use in a service organization. So much for the history, but what do standards, and in particular ISO 9000, actually achieve?

Let us first understand that ISO 9000 exists to give the customer confidence that the product or service being provided will meet certain specified standards of performance and that the product or service will always be consistent with those standards. With organizations in New Zealand who are in the export business, ISO 9000 will often be a prerequisite required by overseas buyers. ISO 9000 gives the buyer confidence.

But what of the organization seeking ISO 9000 certification? Are there any internal benefits? For a start, by adopting a standard such as ISO 9000, the methodology of the ISO system will show an organization how to go about establishing and documenting a quality improvement system. To achieve accreditation, an organization has to prove that every step of the process is documented and that the specifications and check procedures shown in the documentation are always complied with. The recording and documenting of each step are long and tedious jobs. Perhaps the most difficult stage is agreeing on what exactly is the standard procedure.

If an organization does not have a standard way of doing things, trying to document will prove difficult, and many interesting facts will emerge. The act of recording exactly what is happening, and determining what the one set method should be from now on, will in itself be a useful exercise. Wasteful activities should be unearthed, and, hopefully, overall a more efficient method will emerge and be adopted as a standard. Determining a standard does not imply that the most efficient method is being used. The standard adopted only shows that there is a standard method (no matter how inefficient), that the method is written down (documented), and that the written-down method will be used every time. The standard method includes the steps taken in the process and the checks and tests that will be carried out as part of the standard process. This will often require the design of new and increased check procedures and a method of recording that each check or test has been done.

From this it can be seen that the adoption of ISO 9000, rather than the streamlining of an organization, can actually serve to reinforce the supervisor's position. ISO 9000 to this extent can therefore be seen to be contrary to the philosophy of TQM. With TQM, operators are encouraged to do their own checking and to be responsible for getting it right first time. Thus the need for supervision becomes less important or even superfluous. With ISO 9000, the standard method will likely be set by management decree and, once documented, it may be seen as difficult to initiate and to carry through changes. The mechanism of changing documentation will be seen as a barrier to change. ISO tends to be driven from the top down and relies on documentation, checks and tests to achieve a standard, and somewhat bland, level of quality assurance. TQM on the other hand, once established, relies on bottom-up initiatives to keep the impetus of continual improvement. However, as the Deming method of TQM does advocate a stable system from which to advance improvements, the adoption of the ISO 9000 approach will mean that there will be a standard and stable system. To this extent, ISO 9000 will prove a useful base for any organization from which to commence TQM.

Manufacturing companies are now encouraged to incorporate requirements for environmental care to obtain ISO 14000. Perhaps, in the not too distant future, we should be working on ISO 20000 comprising all 20 foundation stones!

Notwithstanding the benefits of obtaining a standard, stable system through ISO procedures, it must be queried why a true quality company should need ISO 9000. Gaining ISO 9000 accreditation is a long and expensive business. Internally it requires much time and effort, and most organizations underestimate the time and effort involved. Generally, recording the systems alone will require the full-time efforts of at least one person. If not required by the customer one must seriously question if the time and effort required to gain ISO 9000 might not be expended elsewhere.

To give an example of how easy it is to underestimate the time required, one small print shop employing 20 people, and with one main contract customer, was advised by consultants that the process would take nine months. To date, one year and nine months have elapsed, and accreditation, although not far away, still has not been achieved. As this print shop has one main customer (90 per cent of the business), and the customer was not pressing for ISO 9000, it is difficult to understand what benefits were expected from accreditation. In fact, the time taken, and the difficulties experienced so far to achieve accreditation, have caused the customer to look more closely at the print shop than it had done previously, and the contract may now be in jeopardy if accreditation is not achieved very soon. What of the expensive consultant? Well, they took their fee and haven't been seen for the last 10 months.

Internal costs are expensive – more expensive than most organizations are prepared to admit. Total internal costs will not be known unless everyone involved in setting up the systems records and costs the time spent. The external costs can be equally expensive. It is not mandatory to hire an external consultant, but there are advantages for doing so. Consultants are not cheap. If you plan to use a consultant, get at least three quotes. Briefing the consultants will force an organization to do some preparatory work which should help in clarifying the overall purpose and give some indication of the effort that will be involved. Once the consultant is employed, it will be your organization that does the work. Consultants merely point the way. They give guidelines and hold meetings, they will help with the planning, but don't expect them to get their hands dirty. They will not actually do any work – you will do the work!

Accreditation can be obtained only through an approved certifying body. The fee charged by the certifying body is relatively small. For the small print shop the accreditation fee was $5,000. Fees depend on the size of the organization and the level of accreditation.

Once the decision has been made to move towards ISO 9000, the next question is which version: ISO 9000, 9001, 9002, 9003 or 9004 (see Figure 7.7)?

ISO 9000 mainly deals with how to choose other ISO series standards for inclusion in a contract between a customer and a supplier.

ISO 9001 should be chosen if there is design work, or changes to designs involved, and/or if after-sales service is required.

ISO 9002 should be chosen if there is no design work involved and/or there is no after-sales service in the contract. Some people think that ISO 9002 is easier to achieve and therefore that ISO 9002 is a lesser 'qualification' than 9001. This is not so. If there is no design work involved or after sales service required, then ISO 9002 is appropriate (it is no less onerous than 9001).

ISO 9003 is the weakest standard. In essence, it requires only final checking and testing of a product before delivery to the customer. Such quality checks are not in tune with the TQM approach. TQM requires everyone in the process to be involved in quality and to be testing and checking at each stage of the process. Reliance on one final check is not a good way of reducing costs of mistakes and of instilling into the organization a quality culture. Of course, ISO 9003 can be amended to include prudent business practices such as quality control of inputs (for example, raw materials), corrective action taken during the process and so on. If such amendments are made, then it would be more sensible to opt for ISO 9002 from the outset.

ISO 9004 extensively uses the word 'should'. This means that an organization is not required to actually do anything included in the standard. ISO 9004 can only be regarded as an advisory introduction paper to quality management. It is not so much that ISO 9004 is wrong in what it covers, it is the lack of compulsion that makes ISO 9004 of little value for contract purposes. If a customer was to use ISO 9004 in a contract document, then, wherever 'should' appears, 'shall' ought to be substituted.

Throughout the ISO 9000 series, reference is made to documentation. The American Society of Mechanical Engineers (1973) defines documentation as 'any written or pictorial information describing, defining, specifying, reporting or certifying activities, requirements, procedures or results'. Documentation is not defined in ISO 9000. But, using the American Society of Mechanical Engineers' definition, a course of action recommended by Sayle (1991), the term allows for 'drawings, diagrams, sketches, VDU displays, software and similar media for communicating information'. Increasingly, organizations are using computers and other forms of information technology such as the Internet and Lotus Notes to communicate and store data. To meet the ISO requirements, it is not necessary to have hard copies of quality plans, quality manuals and procedures. Indeed, when people have a computer terminal at hand, they are more likely to search the computer rather than leaf through large manuals. Also, with a computer system, it is easier to update the records with the latest procedures. By requiring the user to acknowledge receipt of a change on the screen, the quality system can be kept almost instantly updated. A record can be maintained of which users have, or have not, been advised or of who have acknowledged amendments.

The other important aspect of ISO is audits. Audits can be carried out internally and/or by external auditors. Only ISO 9001 and 9002 require quality audits and they require only internal audits. The audit requirements of the ISO 9000 series are more towards compliance checks after an activity has started or been completed. This type of check confirms that procedures are being kept to, or that an outcome complies with the standard. Where mistakes are found, they are after the event rather than before. They will highlight where errors have occurred and thus indicate the need for corrective action for the future, but they do not stop the error happening in the first place. The most effective audit is the audit carried out before an activity occurs, the aim being to prevent mistakes happening. ISO 9000 does not require this type of audit.

To be effective, the internal quality auditor should be trained in audit procedures and the purpose of auditing. Auditors should not regard themselves as police officers. They are not there to set traps or to catch people out. They should be there to help and guide. Obviously if the audit is before, that is preventative, so much the better. It goes without saying that the auditor must know the procedures, understand the product or services, and have a thorough understanding of what is in the contract with the customer. Not all the terms and words used in the ISO series are unambiguous. This is especially true if the ISO is being applied to a service industry. It would seem prudent to agree, between the contracting parties, a definition of terms. If you are getting involved as a quality auditor for ISO, it is strongly recommended that you keep Allan Sayle's book close at hand!

In conclusion, with TQM the aim is continuous improvement and with the continuing impetus for quality improvement being driven from the bottom up. ISO, or the adoption of any other standard, will not achieve this. At best, ISO can be seen as a step on the way to TQM. At worst, it might actually inhibit TQM, as it relies on the setting of top-down standards and controls. A true TQM organization does not need ISO, but, if ISO is insisted on by a customer, it can be made to fit into the overall TQM plan.

Foundation 16: Financial management

Historically the relationship between financial management and manufacturing management has been like oil and water, 'them and us'. The 'quality movement' of the 1980s appeared to have encouraged some manufacturing managers to move away from involvement in costs and measurements. Some manufacturing managers took the stance that cost and measurement were 'internally focused,' the concern of the 'bean counters', whereas the quality movement was externally customer focused. But in fact this was not what the quality gurus such as Deming, Juran, Crosby, Feigenbaum and Peters were saying. Their message was that measurement is important in achieving quality. For a start without a score card of some type it is not possible to determine if improvements are being made (see our comments in Chapter 6 concerning manufacturing performance).

Traditionally accountants have seen themselves as the major channel through which quantitative information flows to management. Accountants work on historical data of what has happened, and their reports cover arbitrarily set periods of time, with little allowance that business activities do not stop on 30 June or 31 December (or whatever other date has been designated as the time to take a snap shot of the financial position of the business). From a conventional point of view, and from the point of view of stakeholders, such as shareholders and bank managers, there has to be a way of measuring the performance of an organization and currently there is no better method than accounting reports. It follows therefore, that for accountants to do their job of reporting to meet the conventional requirements, information will be required from the manufacturing arm of the business. This cannot be disputed. Therefore if information is being provided, then it is useful to try and use that information to improve the productivity of the organization.

It makes sense therefore, that financial factors are integrated with manufacturing and that manufacturing managers can focus on the cost advantage of manufactured goods. Improved quality, delivery and manufacturing flexibility should eventually improve the profit margin, but the impact of any manufacturing cost is straight to the accountants' 'bottom line'. After all manufacturing is responsible for an ex-works cost which accounts for a significant part of the cost of sales. There are indications that there has been a gradual shift in manufacturing towards financial management, probably influenced by the following factors:

- The growth of the 'share owning' population has generated a new breed of consumers who are interested in the financial performance of a company.
- This has required financial management to become conscious of external requirements.
- With the increase in external sourcing and third-party operations, the cost base and its control in manufacturing have been sharpened.
- The economic recession in the late 1980s and early 1990s forced many manufacturing industries to adopt restructuring and cost-reduction initiatives.

It is therefore important for any company to focus on the key issues of financial management in order to enhance competitiveness through manufacturing cost advantages. These issues include achieving business objectives, understanding strategic cost factors and cost effectiveness.

Achieving financial objectives

We do not intend to delve into the sophisticated world of financial management involving the method of financing, tax implications, currency movements, etc. However as indicated earlier it is important that key financial parameters and objectives of the business should be understood and incorporated in manufacturing objectives. Key financial concepts are:

- **Sales value:** The total turnover of the business in money terms.
- **Net profit:** The money made by the business after charging out all costs. This can be expressed before tax or after tax.
- **Capital employed:** Total investment tied up in the business comprising shareholders funds. With the double entry system of accounting, shareholders' funds, or capital, will always equal the total of all the assets less all the liabilities.
- **Working capital:** Working capital refers to the funds available, and is the difference between current assets (debtors, inventory, bank bal-

ances and cash) less current liabilities (creditors, short-term loans and the current portion of long-term loans).
- **Cash flow:** Cash-flow statements show where and how the working capital has increased or decreased.

There are only four basic sources for an increase in working capital and likewise only four basic uses to explain a decrease in working capital, namely:

Increase in working capital

- Profits from operations
- Sale of fixed assets
- Long-term borrowing
- Increase of shareholders' funds through the issue of shares.

Decrease in working capital

- Losses from operations
- Purchase of fixed assets
- Repayment of long-term loans
- Distribution of profits to shareholders (dividends).

The key financial indices influencing the financial objectives of a business are:

Trading margin	=	Net profit \ Sales value x 100
Asset turn	=	Sales value \ Capital employed
Return on investment (ROI)	=	Net profit \ Capital employed x 100

Balance sheet ratios

The main balance sheet ratios are as listed below together with illustrative figures. (The absolute figures are kept to abnormally low amounts so as to avoid a plethora of zeros or the need to continually remember the figures are all in thousands.)

1. Example of balance sheet figures

Capital

Ordinary	8 000	Fixed assets		13 000
Unappropriated profit	2 000	Current assets:		
Equity	10 000	Stock	4 000	
Debentures – 7 per cent	6 000	Debtors	2 000	
Current liabilities	4 000	Cash	1 000	7 000
	20 000			20 000

Sales 6 000; cost of goods sold 4 500, net profit (before tax) 1 200; dividend 10 per cent.

2. Solvency ratios

These show the extent to which a company can meet its current commitments.

$$
\begin{aligned}
\text{Current ratio} &= \text{current assets} \div \text{current liabilities} \\
&= 7{,}000 \div 4{,}000 \\
&= 1.75{:}1 \\
\text{Liquid ratio} &= \text{liquid assets} \div \text{current liabilities} \\
&= 3000 \div 4000 \\
&= 0.75{:}1
\end{aligned}
$$

(Liquid assets = debtors and cash)

3. Equity ratios

These show the extent to which the company is financed by shareholders:

$$
\begin{aligned}
\text{Equity} \div \text{total capital employed} &= 10\,000 \div 20\,000 \\
&= 0.5 \\
\text{Equity} \div (\text{equity} + \text{long-term debt}) &= 10\,000 \div 16\,000 \\
&= 0.625 \\
\text{Equity} \div \text{fixed assets} &= 10\,000 \div 13\,000 \\
&= 0.77
\end{aligned}
$$

(Equity = shareholders' funds including reserves)

4. Operating ratios

These show operating performance in terms of sales and capital employed:

(a) Sales

$$
\begin{aligned}
\text{Capital turnover} &= \text{Sales} \div \text{Total capital employed} \\
&= 6000 \div 20\,000 \\
&= 0.3 \text{ times per year} \\
\text{Equity turnover} &= \text{Sales} \div \text{Equity} \\
&= 6000 \div 10\,000 \\
&= 0.6 \text{ times per year}
\end{aligned}
$$

Current assets turnover = Sales ÷ Current assets
= 6000 ÷ 7000
= 0.86 times per year

Working capital turnover = Sales ÷ Working capital.
(Working capital is sometimes referred to as liquid assets, i.e. normally cash and debtors.)
= 6000 ÷ (7000 − 4000)
= 2.00 times per year

Stock turnover = Cost of goods sold ÷ Stock
= 4,500 ÷ 4,000
= 1.125 times per year
Or in months $\frac{12}{1.125}$ = 10.67 months
i.e. inventory = just on 11 months' usage.

Debtors turnover = Sales ÷ Debtors
= 6000 ÷ 2000
= 3.0 times per year
Or in days
(Debtors × 365) \ Sales
= (2000 × 365) \ 6000
= 121 days to collect sales.

(b) Return

For these examples net profit is before tax. Some organizations use net profit after tax.

Return on investment = Net profit ÷ Total capital employed
= (1200 ÷ 20 000) × 100
= 6%

Return on shareholders funds = Net profit equity
= (1200 ÷ 10 000) × 100
= 12%

Return on sales = (1200 ÷ 6000) × 100
= 20%

Return on working capital = Net profit ÷ Working capital
= (1200 ÷ 3000) × 100
= 40%

Current ratio (current assets ÷ current liabilities)

This is perhaps the most common balance sheet ratio. It indicates the level of safety involved in relying on current assets being sufficient to pay current liabilities. (Note that a company which is forced to sell fixed assets to pay its liabilities is reducing its trading potential and so lessening the profits on which the continued existence of the company depends.) Traditionally, the ratio is supposed to be about 2:1. Too high a ratio suggests inefficient use of capital, for example, excessive debtors or stock, while too low a ratio suggests a lack of liquidity.

Liquid ratio (liquid assets ÷ current liabilities)
When using this ratio, a decision has to be made as to what makes up liquid assets. Cash is a liquid asset and so are debtors, though if the terms of trade allow long-term credit, only immediate debtors may be included. Traditionally a ratio of 1:1 is looked for, but much depends upon how soon each class of liability must be paid; tax payable in nine months is quite different from a nervous creditor whose account was payable some two months previously.

If the company should have an unused portion of an agreed overdraft limit, the unused portion is sometimes regarded as cash because it is available to pay liabilities. That is, the full overdraft limit is listed among the current liabilities, and the unused portion is listed as a cash balance in the current assets. This will improve a ratio that is worse than 1:1.

Equity ratios
For considering equity three ratios can be used.

1. *Equity to total capital employed.* This indicates how much of the total capital being used by the company has in fact been supplied by the shareholders. **Note:** Total capital employed is defined in various ways by various people. Basically, however, it is the balance sheet total, although some authorities exclude intangible assets and some also deduct current liabilities.
2. *Equity to equity plus long-term debt.* This measures the proportion of the permanent capital that is financed by the shareholders.
3. *Equity to fixed assets.* This indicates the extent to which the shareholders have financed the fixed assets.

Operating ratios
The operating ratios can be classed as follows.

1. *Sales to capital.* This ratio measures the efficiency of the use of capital. The higher sales per pound of capital the more effectively is capital being employed.
2. *Cost of sales to stock and sales to debtors.* These ratios help to assess whether stock is too high or debtors are taking too long to pay. Our

example above shows that it would take 11 months at the current rate of sales to sell all the stocks. Similarly, the debtors are taking on average four months to pay. Whether these examples show a poor situation depends on the business and its terms of trade; at face value they would certainly seem to be excessive.

3. *Return on investment and return on sales.* These ratios are widely used as measures of efficiency and performance evaluation. In addition, wide use is made of return on investment to assess the validity of new projects. Most companies set a minimum return on investment rate that must be exceeded before a new project can be proceeded with.

In spite of some recent criticisms, ROI has continued to be the most important single index of the financial objective of a manufacturing business. Hamel and Prahalad (see Chapter 6) attacked managers obsessed with denominators (capital employed). The right approach of manufacturing is, as shown in Figure 7.8, to identify high leverage points of both increasing profits and reducing capital employed. Low-cost manufacturing is a desirable manufacturing objective as long as the investment decisions are geared to longer-term requirements and the measures do not affect the specified standards of quality, delivery and safety. The measures indicated in the ROI improvement tree (Figure 7.8) have been covered in other sections of the book, but it is useful to focus on a total picture of cost advantages so that the inter-relationship between different elements and their relative weight can be visualized.

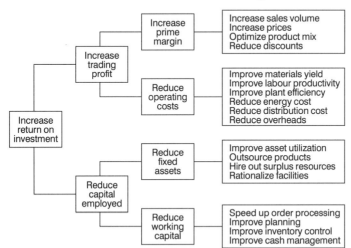

Figure 7.8 *Company profitability: tree of improvement*

132 TOTAL MANUFACTURING SOLUTIONS

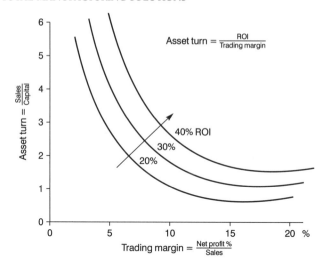

Figure 7.9 *Capital assets productivity*

In special cases, simulation of cost modelling is justifiable.

The financial objectives of a manufacturing business include increasing asset turn, improving profit margin and improving ROI, but these three indices may appear to be conflicting, as shown in Figure 7.9. For a given ROI, profit margin goes down with increased asset turn and vice versa. However, when analysed more closely by managing the improvement of both numerator and denominator (i.e. operations improvement and asset management) the company performance can move to a higher ROI curve and retain improvements in both profit margin and asset turn.

Understanding strategic cost factors

There are a number of strategic factors affecting the manufacturing cost. We shall review two areas, costs determined by volume (including variety and variations) and investment policy. These are strategic in a sense that they relate to the way the company may decide to react to the competition and to developments in the marketplace.

Volume, in general, is good for business as the higher volume reduces the overhead or fixed cost per unit of production. However the advantage of 'economies of scale' should not be pushed beyond the natural capacity of a site as the unit cost could go up due to constraints in site capacity and services. As variety increases, unit cost of manufacturing may also increase due to technology cost, lower utilization of plant and increased overhead/infrastructure. As discussed in Chapter 6, with flexible manufacturing variety can be essential to be competitive in segmented markets. Manufacturing should in these cases accommodate variety by incorporating higher flexibility of plant and operations. Variation is another determinant of product cost. If there are unstable variations in sales demand, supplier lead time, and plant performance, then the planning effectiveness will go down and buffers in stock, capacity and resources will be necessary.

The criteria of investment decision has been covered in Chapter 6. It is important that formal investment appraisal procedures and investment policies are in place. However, the rate of discounted cash flow (DCF) yield should vary according to the type of investment as indicated in the following table.

Table 7.1 *Discounted cash flow yields*

Cost reduction projects	7.25%
Capacity expansion	
Replacement	20–25%
Strategic	15–20%
New technology	10–15%
Environment and safety	0–15%

Evaluation should include all tangible benefits and intangible benefits. The above table is indicative only to demonstrate the relative importance of investments. The actual limit of DCF yield is set by each company depending on financing charges, depreciation rate for a capital asset and the life cycle of the product.

Accounting systems

It is vital that the company has a reliable accounting system in place to provide fast and accurate cost information. The minimum requirements should be standard costing and budgetary control.

As discussed in Chapter 4, some companies are moving towards activity-based costing (ABC), particularly for supply-chain management. The accurate cost information provided by ABC can give a company a competitive advantage. However the experience of western companies according to De Meyer and Ferdows (1990), suggests that the implementation of activity-based costing has not been successful, perhaps due to the historical inertia of standard costing. Any half baked implementation could be more harmful than useful.

Cost effectiveness

Cost cutting or cost reduction exercises, if they are panic driven, or 'chairman's 5 per cent reduction target' will only give short term results and will cause imbalances and disruptions in operations. Other legitimate concerns will be the negative effect on quality, innovation and customer service. And although direct factory labour might account for only 5 to 15 per cent of the total ex works cost (see Figure 7.10) the overwhelming emphasis usually is given to the reduction of labour cost. New and Mayer (1986) state,

> Whole work study departments are maintained to control the direct labour content of unit cost. Yet there are many plants that spend twice as much on purchased materials as on direct labour that do not even attempt to measure purchasing performance realistically.

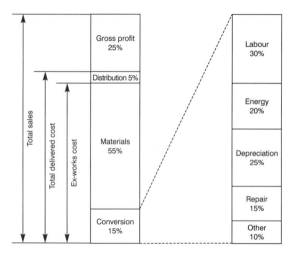

Figure 7.10 *An example of a cost structure*

The real business focus should be to survive and to be capable of competing in the future. Although strategy and innovation are important, the hard fact is that unless there is a positive operational cash flow the business cannot plan for the future. Therefore it is vital to have cost improvement even in a profitable company, but the approach must be cost effectiveness, not cost cutting.

The key principles of a cost effectiveness programme are:

- Understand the strategic drivers of cost i.e. volume/capacity, variety and variation and their impact in the marketplace and competition.

- Evaluate the effect of any company saving measures on quality, safety and customer service.
- Identify the leverage of cost structure and set priorities of effort and 'go for gold.' But as a rough guide the amount of effort allocated to manufacturing cost reduction should be proportional to the rest of the costs of the company.
- The programme should be company wide, although manufacturing is the key player.
- The principles of value engineering should be applied in identifying and emphasizing operations where no value is added.

Cost effectiveness is a continuous process for all manufacturing businesses, but some businesses may require a quick and significant change in their cost structure. Study teams should then be formed to carry out ad hoc exercises, such as big scale value analysis, restructuring or site rationalization (including plant closure).

Value analysis is a technique used to examine each element of a process so as to find a cheaper material or better method with the aim of maintaining or enhancing the value of the product in performance terms and at the same time reducing the cost. Big scale value analysis (BSVA) uses value analysis technique but in addition examines the total delivered cost (see Figure 7.10) of the business and has a short time scale (usually less than one year) with emphasis on company-wide implementation. The cost model in Figure 7.10 is a typical example of an FMCG business and obviously the proportion of cost elements would vary depending on the product.

Foundation 17: Information technology and systems

Information technology (IT) is rapidly changing and becoming more powerful. It will be a continuing source of competitive advantages for manufacturers if used correctly. In 1995 the personal computer (PC) on the desk of an average manufacturing manager has the capability of 10 to 12 mips (millions of instructions per second) of computing power, 4 to 16 megabytes of main memory and 300 to 500 megabytes of direct access storage. Ignoring the technical jargon, most of us have on our desks more computing power than the average £100 million a year manufacturing plant had 10 years ago. This IT revolution is available to everyone and how a company puts it to work will determine to a great extent its competitiveness in the global market.

The rapid growth of information technology has also created problems and challenges. Many senior managers of companies lack any detailed understanding of the complexity of technology. They either follow the fashion (e.g. 'no one was fired for choosing IBM') or they are discouraged by the cost of technology, or from a lack of evidence of savings in a new field. When executives read about all the clever things seemingly low cost computer technology can do they feel frustrated when the systems experts say, 'It will take three years to develop the software'. Most senior managers also feel lost in a blizzard of buzz words.

Another problem is that, until the late 1980s, the lack of protocol or standard method of communication between systems created isolated systems and 'islands of automation'. A CBI Report on information technology says,

> General Motors, the world's largest manufacturer, has over 40 000 intelligent machines in use for design and manufacturing operations. GM estimates, however, that only 15 per cent of these machines can communicate with other systems and that up to 50 per cent of the cost of installing new automation is spent on facilitating one computer to speak to another.
>
> Information Technology, CBI Report, UK

Incompatible information systems usually result from looking at bits and pieces of processes to solve an immediate problem. It is no consolation to know that it is not just manufacturing people who get stuck with an 'island of automation' that keeps everything to itself. For example, a distance learning education institution that we know of, standardized on Apple Mac word processors for the production department (some 20 word processing staff) and provided the tutors and writers with IBM compatible PCs. Certainly disks produced by the writers can be translated for use on the Apple Macs, but this takes time and difficulties in formatting do occur. Such problems can be prevented by serious planning and adopting an 'open system'. IT hardware strategy will be discussed later.

Another issue is software application strategy and networking. Successful companies are using information technology and compatible software on the factory floor (CAD/CAM and process computing), administration, planning and in the office (word processing, spreadsheets and PC work stations). Sophisticated global information networks (E-mail, the Internet, Lotus Notes and high speed data and video links) have simplified international operations. Electronic data interchange (EDI) technology has made possible extended supply chains between companies, for example cash register transactions can trigger automatic reordering. On the down side there are also many failures when a company's software does not meet its requirements or wasted effort when IT managers try to re-invent the wheel by attempting to build software in-house when simple tested 'off-the-shelf' software would suffice.

Yet another issue is the implementation of systems to the benefit of the users. When a company looks for an IT solution to a problem without re-engineering the process, refining the existing database or training the end users, the application is doomed to fail. Real disasters can be very expensive. For example the $60 million Master Trust accounting system for Bank of America had to be scrapped because it could not keep accurate accounts.

Figure 7.11 shows a framework of IT strategy comprising three levels – hardware strategy, software strategy and implementation strategy.

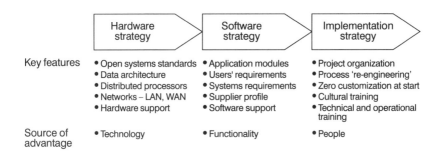

Figure 7.11 *Information technology strategy*

IT hardware strategy

Dramatic developments towards 'open systems' standards started in 1987 when AT&T in partnership with Sun Microsystems introduced the Unix Open Look operating system. This system was used by Wang, Oracle, Olivetti and Lucky Gold Star. Seven big computer companies led by IBM, Hewlett-Packard and DEC formed the Open Software Foundation (OSF) and introduced in 1990 their own competing standard operating system using IBM operating systems as core technology. Fortunately the interface standards of both competing standard operating systems do not differ significantly. However should there be a further polarisation of the two camps it is possible that the majority of companies will follow the Open Software Foundation Application Environment Specifications (OSF AES) which operate IBM-AIX, DEC-Ultrix and HP-UX operating systems.

There are good open standards such as ANSI 92 for a relational database system which conforms to the SQL standards. The leading proprietary database systems which conform to these standards include Oracle, Sybase, Informix and Progress. In order to ensure the maximum level of portability, the future direction of new software is likely to move towards the so called three-tier architecture. For example Tier I contains the user interface, Tier II is the functionality layer and Tier III is the database layer.

With the rapid development of application tools a proven hardware policy has been what is known as client-server computing. All 'servers' are open system large or mini computers (e.g. IBM-AIX) and 'client' computers are largely personal computers (PCs).

The benefits of standards include the creation of local area networks (LANs) and wide area networks (WANs). A LAN can cover a large industrial complex while a WAN can offer inter-site communications on a national or international basis. In the early 1990s the companies were gradually migrating from previously popular network standards (such as PC LAN, Novell, Internet) to open systems network such as NFS-based systems. However by mid-1009s Novell started to regain the market dominance.

The hardware strategy should also include the capability of local hardware support both by suppliers and the company's own staff. The support capability may influence the selection of hardware whether IBM, HP, DEC or SUN or other. A sensible strategy is to go with the market leaders who are setting the de facto standards.

IT software strategy

At the early stage of information technology, applications software was limited to financial and commercial areas. Now a company is faced with a bewildering array of software ranging from design/process engineering, to manufacturing, to supply chain, to administration. Versions of specific software and systems technology will continue to change. Therefore it is vital that a manufacturing company formulates a software strategy by careful planning.

The first step is to identify the areas of application depending on the activities size and priorities of the company. Figure 7.12 shows a framework of application software in five key areas, namely financial administration, supply chain management, factory administration, and 'client' work station. The traditional computing modules of accounts and payroll are in financial management. The biggest area of application is in supply chain management starting from sales forecasting to customer service and electronic data interchange (EDI). At the factory shop floor there are two application areas, namely factory administration – comprising management information systems – and factory automation – comprising design, process engineering and automation of equipment. The software for client work stations is PC based (usually supplied by two large global suppliers – Microsoft and Lotus) covering word processing, (e.g. Wordpro, Word for Windows) spreadsheets, (e.g. Lotus 1-2-3, Excel) computer graphics (e.g. Freelance), multi-tasking (e.g. Windows 95), E-mail and conference (e.g. Lotus Notes). During the late 1980s many manufacturing companies searched for one turnkey package and invested in what is known as computer-integrated manufacturing (CIM) with limited success. If a company follows an 'open systems' policy for hardware and relational database then different proprietary software packages stand a better chance of being interfaced and database information can be shared in a client-server environment.

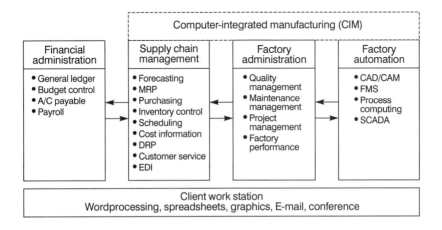

Figure 7.12 *Application software modules*

The software policy should include standard packages for the company in specific areas of application. The selection of software should conform to the key criteria of user requirements, systems requirements, supplier profile and software support. The earlier examples of applications software were relatively inflexible and the approach was 'systematize the customer' rather than 'customize the system'. Many disillusioned customers attempted to build their own software and burnt their fingers in the process. In the present climate the software tools have become flexible, the IT technology is advancing rapidly, competitive expert support is provided by specialist software houses and thus it is prudent to buy appropriate software rather than to develop your own (see Figure 7.13). The software should conform to open systems requirements and the supplier should be both reputable and locally available for support. The company should also build up its own IT support staff, especially a 'user support' service.

140   TOTAL MANUFACTURING SOLUTIONS

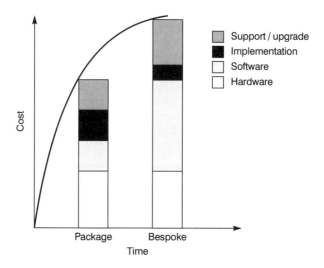

Figure 7.13 *Software development strategy*

Implementation strategy

The success of an IT strategy depends as much on the selection of appropriate hardware and software as on their implementation.

Similar to a company-wide programme such as TQM, the implementation must have top management commitment. This should be reflected in setting up a project team comprising members from users (marketing, logistics, manufacturing, accounts) and business systems. The project manager is usually chosen from the main user group. For example if the application software is for supply chain management then the project manager should ideally have a logistics background.

The project team should receive both technical training (e.g. Unix, Oracle) and operational training (functionality of the software). The project manager then prepares a clearly stated action plan with target dates and resources for key activities. The plan must include review points and steering by the members of the board.

It is essential that the existing procedures and processes are thoroughly and systematically reviewed. There are various tools for analysing the flow and requirements of the existing systems. Statistical process control (SPC) techniques are widely used. Nowadays some companies are using computer-aided software engineering (CASE) tools to analyse the structure, database and flows of the existing process and compare them with the proposed software for implementation. With the success of the business process re-engineering (BPR) approach of

Michael Hammer and James Champy (1993), some companies are using an IT application as a catalyst and applying the principles of BPR to re-engineer the total business processes of the company. The approach should depend on the depth and breadth of the application systems, but there is no doubt that the existing systems must be reviewed and refined when implementing a new system.

One important rule is that the user should not try to customize the system at the outset. Often after acquiring experience on the new system the user may find that the need and nature of customization could be different. However it is necessary that a 'prototype' is tested for a new system using the company's own data.

After the training of the project team the training programmes should be extended to all potential users of the system. The training features should contain both cultural education to establish acceptance by everyone concerned and operational training to understand the functionality and operations of the new system. Training documents are designed specifically for the users' needs. The next stage is the data input and 'dry run' of the new system in parallel with the existing system before the system goes live. There are benefits of forming users' group for exchanging experience with users drawn from within and from outside the company.

Summary and further thoughts

This chapter has covered quality management, financial management and information technology. We are not suggesting that we have written a complete accounting text book or the definitive work on information technology. Far from it.

With quality management we have however gone into greater depth. Quality management is not a discipline restricted to one body of knowledge or expertise. Quality management is for everyone in the company to know and to understand in detail.

We have shown quality management has two levels – basic requirements of specification, time and cost, and higher level requirements covering service and customer focus issues. We accept quality has a price but the cost of not performing can be unknown and is probably unknowable.

We also discussed a hierarchy of quality methods ranging from inspection at the end of the process, to no inspection by supervisors and the reliance on suppliers, and each worker in the process to get it right first time, every time. For such a bold approach to be viable – e.g. no supervisors, no inspectors – workers must be empowered. But more than that, they must want to be empowered, and managers must believe and trust. For most companies this is a desirable goal but probably not something to be attempted overnight!

We also covered in some depth ISO 9000. We believe that many people see ISO 9000 as a goal in itself. We say ISO 9000 may be a step on the way to TQM but it is only that – a small and expensive step. We suggest that a true TQM company does not need ISO 9000.

With financial management we introduce key concepts and ratios. Unless the factory manager understands these ratios he or she will always be at the mercy of the accountants. The ratios are explained simply, and illustrated with easily understood examples. If you have some accounting knowledge, don't skip this section, take five minutes to work through the examples and consider how they apply to your organization.

Some time is spent on ROI and some time on cost cutting. Both these areas are of particular concern to the factory manager. ROI can be used to prevent you getting much needed equipment. Cost cutting, if applied 5 per cent across the board, will inevitably hit the factory the hardest. Other sections probably do have some slack or spare capacity – but does your factory? It is important that the factory manager understands ROI and that the factory manager can defend him or herself against ill judged cost cutting exercises.

For information technology we have taken a more general approach. This section is equally applicable to all functions of the organization. The key issue in any new IT system is knowing what you want, going with a system with local support, and initially making do with off-the-shelf software. We have not discussed uninterrupted power supply, disaster recovery, the need to back-up files and so on. All these issues are nuts and bolts and should be second nature to your IT manager. This section was not written for the professional IT manager. It was written to give the average manager an understanding of the strategy of IT implementation.

References

De Meyer, A. and Ferdows, K. (1990) *Removing the Barriers in Manufacturing.* INSEAD.

Hamel, G. and Prahalad, E.K. (1994) Competing for the Future. *Harvard Business Review*, Jul-Aug, 72, No. 4.

Hammer, M. and Champy, J. (1993) *Reengineering the Corporation.* HarperCollins.

New, C.C. and Mayer, A. (1986) *Managing Manufacturing Operations in the UK.* British Institute of Management.

Oakland, J.S. (1992) *Total Quality Management*, Butterworth-Heinemann, Oxford.

Sayle, A.J. (1991) Meeting ISO 9000 in a TQM World. AJSL, UK.

Taiichi, O. (1988) *Toyota Production System.* Productivity Press.

Taylor, F.W. (1947) *The Principles of Scientific Management.* Harper & Brothers, New York.

8

People

> *How beauteous mankind is!*
> *O brave new world,*
> *That has such people in't!*
>   William Shakespeare
>   *(The Tempest)*

The last, but definitely not the least, pillar of total manufacturing solutions is people.

A manufacturing company becomes a winner over its competitors by attaining high performance with best practices and by sustaining it through the efforts of its people. Certainly innnovation is a competitive weapon, but new product designs and formulations can be copied, and usually sooner than later. Likewise new technology and materials are procurable by competitors on an open market. The success factor for attaining and sustaining a high performance which will distinguish one organization from another will be the quality of its people. People power includes quality of and consistency of performance.

In most companies the function once known as personnel has been remodelled as 'human resources management'. The change in name signals a change in emphasis. Human resources managers of today deal with issues more far reaching and complex than those previously handled by personnel managers. Apart from having to understand, and advise on, continuously changing employment and workplace related legislation (such as health and safety, employment contracts, and so on), organization structures have changed, from the old hierarchical organization chart model where every one knew their place, to 'borderless' organizations with flexible work practices and the need for continuous retraining and education. The implication of these challenges is being experienced by all levels of management in every function. Life may have seemed simpler in the 1960s and 70s but one should not forget the disastrous effects of poor industrial relations of those years.

The study of human resources and industrial psychology is a fascinating field. Capital assets depreciate with time while human resources can appreciate with experience and appropriate learning programmes. It may sound a cliché but knowledge, skills and initiative of employees are a company's most valuable asset. The challenge with knowledge is that it is more difficult to manage than capital. Knowledge is stored in the heads of employees rather than stocked in a bank, warehouse or in a machine. Knowledge takes time and money to acquire and it should not be lightly disposed of with restructuring and the loss of skilled people. Properly encouraged people can learn new skills.

We have included three final foundation stones related to the management of people, namely:

18. Management skills and culture: This covers the structure of the organization, the culture of the organization (how we do things around here), calibre of the management team, leadership and flexibility, team working, and empowerment.
19. Flexible working practices: This includes a defined policy to attract highly skilled employees, the development of workers, flexibility in work systems and reward systems.
20. Continuous learning: Nothing is static in today's world. Technology is changing daily. Training and educational programmes with appropriate resources and facilities are essential at all levels of the organization.

Foundation 18: Management skills and organization

Both academics and practitioners agree that the major obstacle to the implementation of a change programme, such as total quality management, is the structure and culture of the organization. A bureaucratic, top-down structure will result in a culture where the management does the thinking, makes the plans and gives orders, and the people on the factory floor will be expected to wield the screwdrivers and to obey the orders. The culture will be such that the lower echelons will not be expected or encouraged to make suggestions.

If an organization is serious about harnessing the power of the people then the structure will have to be such that people at all levels are encouraged to make suggestions, and that communication will become multi-directional, that is up and down the 'chain of command' and across functional and departmental borders.

In a customer-focused business manufacturing managers and logistics managers in particular must be adaptable to the needs for organizational change. They are the people closest to the factory operatives on whom the quality of the product depends.

The organization's culture, missions and commitments begin at the top of the organization. It is essential that everyone should clearly know what is expected and what is acceptable. Top management should be seen and known at least by sight at all levels of the organization. Top management has to be visible and has to lead by example.

It is expected that the management team will consist of high calibre professionals with people skills. Management however has to be continuously appraising their own performance and the management structure.

Too often the attitude is that total quality management can be achieved by issuing a new mission statement that says all the right things such as 'people are our greatest resource', and by introducing flexibility, team work and empowerment in the factory. The philosophy is that quality of product is a factory issue, and there is no need to change the structure of the rest of the organization.

Mission statements and organizational culture are further discussed in Chapter 11. In this chapter we discuss organizational structure.

Organizational structure

The traditional organization structure followed by many manufacturing companies is based on functions (see Figure 8.1). With this type of structure we are likely to find divisions such as financial, marketing, technical and personnel. Within each division we might find responsibilities by function as an accounts department, a purchasing department, a planning department and so on. Typically each function or department is driven by a budget. Each manager, or functional head, guards their 'territories' from other functional barons. Their prime concern is often to keep expenses within budget. In addition to the waste of effort in internal 'in-fighting' and in 'empire building' the issue becomes serious when a customer seeking some information is passed from one department to another, with no department prepared to accept responsibility. We discussed some problems of traditional organizational structures in Chapters 3 and 4.

The leading-edge companies, realizing that the main objective of business is to create added value outputs and not inputs, are moving towards 'flat' organizations. There is however a danger of adopting the latest organizational 'fad' without thinking through what is appropriate for the business and the steps required to change the existing team. Another approach is the matrix approach where people work for one department but have responsibility for a project in another department. This approach tries to superimpose a cross-functional team onto a rigid

146 TOTAL MANUFACTURING SOLUTIONS

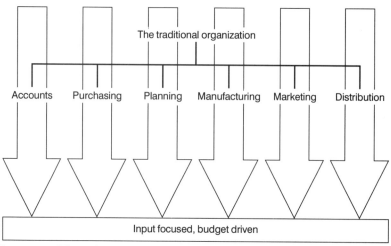

Figure 8.1 *The traditional organization*

budget-driven departmental structure. The limitations of a 'matrix organization' are readily apparent as depicted in Figure 8.2. (For example the confusion that arises with two or more bosses during a project or a decision process and the conflicting requirements and priorities for individual members.)

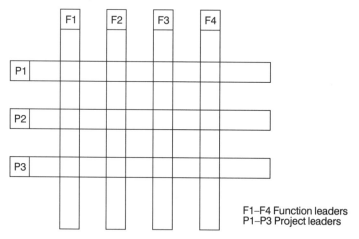

Figure 8.2 *The matrix organization*

One radical approach to organizational changes is the borderless organization as shown in Figure 8.3. This is particularly effective in supply chain management. In a borderless organization the focus is on the customer and therefore the organization is market driven, rather than budget driven, and responsibility for dealing with a customer's query is the responsibility of everyone.

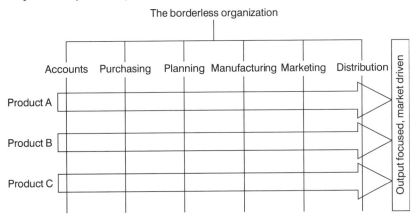

Figure 8.3 *The borderless organization*

It is easier to modify a piece of equipment from what is shown on an engineering drawing than it is to change an organizational structure from what is drawn on a piece of paper. Organizations tend to grow around the existing organizational chart rather than taking a clean slate approach. Ideally organizational changes should follow a holistic approach including identifying resistance, conflict resolution, listening, feedback and communication planning. Kolodny (1993) suggests a 'step by step road map' for the implementation of significant organizational changes (see also Carnall, 1993).

Management profile

Management of people in manufacturing has always been important, but in the 1990s it has become even more so due to global competition, shorter product life cycles, rapid introduction of new technology and the impact of corporate programmes (such as total quality management and business process re-engineering). We have discussed the need for the whole of an organization to be involved in a culture change to total quality and not just the factory. None the less the factory is central to the improvement process.

If one compares the companies that have excelled, either in performance or in a quality programme, with those that haven't, the difference is in the strength of manufacturing management. All organizations have access to the same books, approaches, consultants and seminars. Similarly the improvement possible through new manufacturing and information technologies is available to all. The failure of implementation lies with attempting to limit a change to one part of the organization or by the calibre of existing managers. No matter how good the implementation programme is, unless the factory managers are of sufficient calibre, world class manufacturing will not be achieved.

We have earlier stressed that managers who are functionally orientated and/or occupied with short-term results and quick fixes are almost guaranteed to produce failure for their organization and thereby for themselves. A manager who can convert manufacturing into a world class competitor has to be multi-functional, intellectually adaptable, willing to take on strategic changes, and to possess both functional skills and an aptitude for general management. It will not be an easy task to acquire these competencies from existing manufacturing managers. However, before we rush off and dispose of experienced 'old' style managers it is worth giving them a chance to change. Human beings can be very adaptable if given the right encouragement. Old dogs can learn new tricks, but they have to know why the new tricks are needed and be willing to adapt. Winning them over will be the hardest part.

In the long run any organization has to think of the future. To train a new manufacturing manager will take some years. A leading edge company will need to develop a new breed of manufacturing manager and will need to plan a flow of young new generation managers.

The new generation senior manufacturing manager will have a manufacturing background and, to gain the benefits of a borderless organization, will have a comfortable interface with every business function, marketing, finance, logistics and personnel. A survey by De Meyer and Ferdows (1990) confirmed that five out of the top ten manufacturing action plans relate to integration. A model for developing such a demanding management profile is to recruit high calibre technical graduates into a well structured corporate management training scheme. For the first two years the trainee should be employed in a meaningful role in one of the core value-added functions of purchasing, logistics, or in a manufacturing unit. From our experience little benefit will be gained by moving a trainee every two months or so from department to department. Trainees need to be able to feel that they are contributing; they will be keen, young, well educated and adaptable, and in high demand. There is no sensible reason as to why they should not be able to provide a useful contribution right from the start.

After the initial two years the trainee can be appointed to an industrial engineer's position, or something similar. Industrial engineers, in this sense, are not traditional 'time and motion study people', but change agents in company-wide programmes such as TQM, BPR and MRP II implementation. After two or three years as an industrial engineer, the new generation manager can then move to other management roles in marketing, manufacturing, or supply chain management. To become a fully rounded manager, after between five to ten years, sponsorship through a part time MBA programme such as offered by the Henley Management College should be considered.

In the past chief executives have had an accounting background, and in some cases marketing backgrounds. In the future the chief executive is more likely to hold an engineering or science degree, supported by a post-graduate qualification such as an MBA from a specialist management studies institution.

Obstacles in organization

In spite of discussions about departments working together to satisfy internal customers and the end users (the real customers), in practice there are often strong personal reasons for resistance to change including power, paradox, perplexity and paradigm (the four Ps of change).

Power

A recent survey by the Hay Group UK (1995) suggests that senior managers are consumed by power and that they only pay lip service to change. Usually it is more important for the next promotion of a senior manager to show how well he or she has run their empire, met the budget, and satisfied their boss than it is to co-operate with other managers. Buzz words are learnt, the right motions are gone through, good reasons are found as to why a change programme failed and the achievements of other objectives are highlighted in annual appraisal reviews. If the person who eventually gets to the top is the best at playing these internally focused games, such a person is unlikely to be able to lead an organization in global competition.

Paradox

It is an apparent paradox that two contradictory concepts can both be true in a company. For example, in a multiproduct manufacturing business one product group is promoting employee empowerment which is totally different from the business process re-engineering projects of another product group. One group is advocating asset utilization for its high-volume products while the other group is striving for flexibility for its large variety of products. The reality is that both camps are right but that they are not moving in the same direction, thus a sense of polarization is created within the company. To get people moving in the same direction, leadership from the top, a common goal, and the means of self-assessment are necessary for lower level staff to feel some sense of belonging or ownership.

Perplexity

Another problem is the perplexity of the change of direction and nature of work that people encounter in business. Change is a dynamic process which requires continuous review, but if the change process is not properly handled people get confused. For example, if a company is trying to rebuild an organization, but before the change is halfway through, senior management decide to launch another programme, and thus the previous work is of little value.

It is said that people prefer not to change. That they are more comfortable with what they know. But in reality people of the 1990s are very used to change and very adaptable. Never have so many people travelled abroad for pleasure (or had the wherewithal to do so) as have this generation. How quickly most of us have adapted to cell phones, personal computers, and concepts such as virtual reality. It is not change which perplexes people, it is the reason for change. If change has to occur then the reason has to be carefully explained, and the change process has to be carefully planned so as to involve all members of the organization. Unless the members of the organization understand the purpose of the change, then change will be difficult to effect.

Paradigm

Psychologically, a change process is too much to handle for many managers who like to follow the chartered routes of traditions. The paradigm is why bother to change if it worked in the past. It is difficult to break out of the mould and many managers once they get to a senior position may pick up the styles, premises and approaches of their predecessors. Privileges of rank such as the large corner office, the cocktail cabinet, the reserved parking space will be clung to. Partly this will be the new executive's own vanity and partly it will be to satisfy the expectations of friends and family. There is one very successful world class organization that we visited where the chief executive does not have an office or a reserved parking space. He doesn't need a parking space, he is always first to work in the morning. He has a desk but he shares a large office with executives responsible for finance, marketing, planning and manufacturing. Thus team work is not only espoused by senior management it is openly practised.

In order to remove organizational obstacles we need effective leadership, appraisal of managers beyond budget performance and a continuous learning process. There is a fundamental need for a paradigm shift in many organizations, for managers to move from being managers to becoming leaders.

Leadership

The style and quality of leadership must come from the top and then filter through to managers and subsequently down to empowered employees. Alfred Sloan, chief executive of General Motors from 1923, did for management what Henry Ford did for the shop floor. Sloan set three clear objectives for managers:

- to determine a company's strategy,
- to design its structure, and
- to select its control systems.

Sloan's approach, particularly to the design of a rigid structure and control systems relying on supervision, is now vulnerable in global competition. None the less the logic of designing a structure and control system to meet the strategy, rather than first establishing a structure and then trying to develop a strategy independent of the structure, is difficult to refute.

The shift is now from a boss approach to a leadership approach. For example the boss controls staff while the leader inspires them (e.g. Richard Branson of Virgin). The boss depends on authority and the leader depends on goodwill. The boss shows who is wrong. The leader shows what is wrong. However even 'inspirational' leadership requires perspiration too. Charisma alone does not bring about results in a business. Academics and psychologists of 'human engineering' believe that the leadership styles (e.g. autocratic, consultative, participative and democratic) should be adaptive to the idiosyncrasies of human behaviours. There is no one ideal leadership style.

Management gurus have suggested two essential qualities of leadership. One is industry foresight and vision, the other is innovation. Hamel and Prahalad (1994) clearly established that 'top managers must recognize that the real focus for their companies is the opportunity to compete for the future'. The other quality is related to the first and involves the innovative attitude to encourage and convert crude ideas into results. As expressed by Drucker (1969):

> A top management that believes its job is to sit in judgment will inevitably veto new ideas ... Only a top management that sees its control function as trying to convert into purposeful action the half baked idea for something new will actually make its organization – whether com-

pany, university, laboratory or hospital – capable of genuine innovation and self-renewal.

Christopher Bartlett of the Harvard Business School and Sumantra Ghosal of the London Business School (1994) focus on two other 'post Sloanist' processes of leadership. The first is 'competence building' by linking up the superior depth and breadth of all the employees' talents, and the second process is entrepreneurism through letting loose independent spirits through small business units. (Bartlett and Ghosal cite successful companies such as Kao, 3M, Canon and Intend.)

Foundation 19: Flexible working practices

Traditionally engineering craftsmen held positions of skill and status in manufacturing industries. These people's positions represented extensive apprenticeship training, a membership of a powerful trade union and an affiliation to a deep rooted culture and tradition. In order to protect their specialized skills and jobs craftsmen formed a large number of trade unions. For example, in the UK in the 1960s, there were different trade unions for tinsmiths, sheet metal workers and boilermakers depending on the thickness of the metal plates upon which they worked. The politics of multiple unions in the same factory resulted in unreasonable demarcation practices and high wages relative to other groups of workers. A job that could be done by one craftsman in fact required three or four plus their mates. In the ship building industry for example, the companies with restrictive union-driven practices priced themselves out of competition. What was intended as a means of job protection turned out to be a job loser. The old single and narrow skill-based approach of traditional engineering craftsmen has become increasingly outdated and inappropriate.

With the changes in manufacturing technology, particularly in process industries, the demand on the technical skill level of production operators has significantly increased and given rise to broadly common engineering craft jobs. Some of the job titles are engineering crafts person or process technician. An appropriate recruitment and development plan for both engineering crafts people and process technicians can be accommodated in the same career development structure. Figure 8.4 shows a model to illustrate the stages of development for the flexibility of engineering crafts people.

The principles of the Japanese lean production system and of total productive maintenance clearly demonstrate that great improvements in both productivity and quality can be achieved by team building and empowering a well-trained workforce. Leading edge companies who

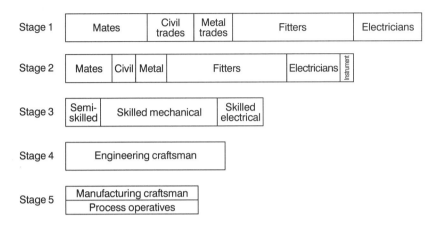

Figure 8.4 *Crafts flexibility*

have successfully applied flexible working practices have achieved, by a varying degree, a reduction in workforce and an increase in operational efficiency.

A detailed study by Cross (1985) showed that engineering flexibility has been considered by all major employers in the UK for a number of years, and that some, such as the Fawley Refinery of Esso, achieved flexibility by productivity agreements and development programmes. The situation has changed significantly over the last 10 years. In the UK many trade unions have merged, and as job losses mounted both employers and unions have realized the importance of flexible working practices in the current economic climate and with international competition. The industrial relations system which exists in factories is conducive to implementing flexible working practices. However there are still instances of confusion, lip service and lack of a systematic approach to manage organization changes. The key areas to be addressed are flexibility of structure, team structuring, industrial relations and continuous training.

Skills flexibility structure

A company needs a clear policy regarding the structure of skills flexibility. The issues mostly revolve around the flexibility within engineering skills (e.g. between mechanical and electrical) and the flexibility between production and maintenance. The key principles of skills flexibility should include:

- Defining job categories to accommodate a common career development plan covering both engineering and production.
- Merging of maintenance and production for shift operations.
- Understanding the importance of specialist technical skills.

The titles and categories of factory jobs can vary depending on the location, language and available skills. However the following model illustrates a representative skill path from which a specific structure can be developed.

Skill path model

Level 1: Operators to be responsible for machine operations, machine setting adjustments, simple change-overs, cleaning, lubrication, simple inspections and preventative maintenance tasks on one or more machines under their control. Machine operations include knowing the standards of quality required and the ability and authority to take corrective action to achieve set standards.

Level 2: (a) Crafts people to be able to (as assigned) carry out corrective maintenance, major change-over, equipment settings, fault diagnostics, inspection and preventative maintenance tasks as well as normal machine operations as described for Level 1. (b) Team leaders will have at least the same knowledge as the crafts people but will have additional manufacturing responsibilities such as achievement of output targets to a desired quality standard. Other responsibilities will include organization of the team and the organizing of resources needed by the team to do their jobs (cleaning materials, oils and lubricants, tools, spares, etc.).

Level 3: Technicians will be able to repair units, and will have specialist skills in mechanical, electrical and ICA (instrumentation, control and automation). They will be able to assist with Level 2 predictive maintenance analysis, and be involved in long-term trouble shooting and the design of processes/systems for trouble-free machines.

Level 4: Engineering or manufacturing supervisors/managers will have higher technical education and training to solve complex longer-term problems, develop maintenance strategies, advise production on optimum line speed, and to provide on the job training support to other levels.
The traditional electrician will no longer be required. In Level 2 crafts people will be trained in basic electrical skills. The specific electrical and ICA know how will be at Level 3.

Factories will apply a high degree of flexibility depending on the availability of skill level. The level of demarcation – should there be any at all – would be based on the capability and skills of employees, not on their job titles or union cards. The recruitment, training and development for each level will be covered in foundation stone 20, training and continuous learning.

It is important that before a task is assigned to a particular level, all the elements of the task are analysed to determine the skill overlap as illustrated in Figure 8.5. Multi-skilling can be an exciting concept but it can present a potential risk in safety and quality if the details of the job and skills are not properly matched: 'a little knowledge is a dangerous thing'.

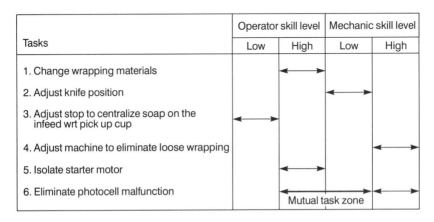

Figure 8.5 *Task sharing - ACMA 711 Packing Machines*

Skills flexibility can only be effective when the long-standing barriers between maintenance and production are broken down to form an integrated manufacturing operation. The first step is to implement a 'manufacturing based' organization as shown in Figure 8.6. In this model the maintenance engineer (area engineer) and his team are physically located in the production area. The area engineer reports to the manufacturing manager and also retains a functional link with the factory chief engineer regarding engineering policy. The next step is an integrated organization where the manufacturing manager is responsible for the reliability of his assets and maintains an advisory link with the chief engineer. This is illustrated in Figure 8.7. There will continue to be a role for the chief engineer with factory services, projects, planned or expert maintenance, planning the installation of new systems, and technical training. Overall the number of levels will reduce, depending on the available skills, and the jobs of supervisors and technicians in both maintenance and production will merge.

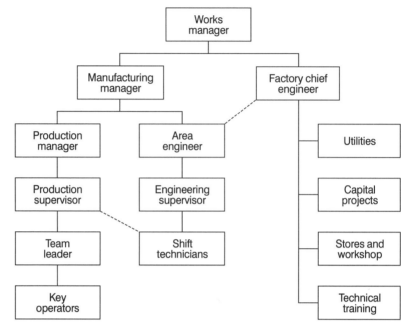

Figure 8.6 *Manufacturing-based organization*

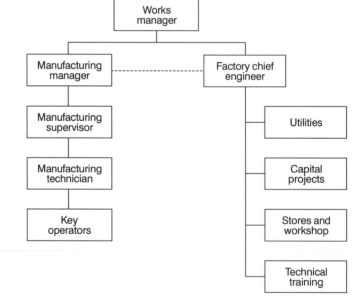

Figure 8.7 *Integrated organization*

Another important issue is the recognition of the need for high level skills, technical expertise, and the special aptitude required in a high-tech operation or high-tech maintenance. The model in Figure 8.8 illustrates the skill levels in a high-speed packaging machine. The depth of knowledge required at the expert level is so diverse for mechanical, electrical and control engineering that it may be prudent not to seek flexibility at these levels. An instrument technician may be an expert in programmable logic control (PLC), but will not necessarily be adept at diagnosing the faults in a 'Geneva' mechanism. Using the analogy that a multiskilled craftsman is a general practitioner, then an expert technician in a particular field is a neurologist.

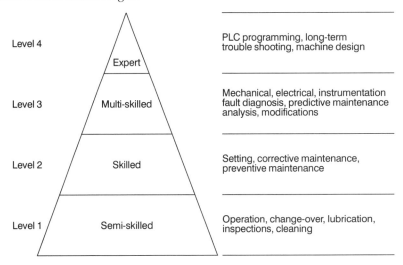

Figure 8.8 *Skill profile in a high-speed packing line*

Team structuring

The traditional concepts of manual flow-line manufacturing and division of labour as detailed in Adam Smith's pin making example (1766), the repetitive short cycle progressive assembly line of Henry Ford and the standardization of work by time and motion studies of Taylor and the Gilbreths have been challenged, particularly in the last two decades. The tedium of assembly line work has been well documented and analysed by R. Wild (1975), and illustrated in the classic Charlie Chaplin movie *Modern Times*. The monotony of short-cycle work has been further aggravated by a number of factors, including the higher educational attainment of workers, the increased unemployment benefits of the labour force, the increased official concern over RSI or OOS (repetitive strain injury or occupational overuse syndrome) and the impact of the 'quality movement'. The experience of Japan's quality circles, and Kaizen, where volunteers of workers form teams to improve safety, manufacturing, or quality problems have proved successful in increasing and sustaining productivity. With Kaizen, companies involve workers in teams with designated team leaders by allowing them a good deal of autonomy in planning and managing daily activities. However what is not often understood is that with the Japanese model empowerment only goes so far. Teams are encouraged to look for new systems and to suggest improvements, but they are not empowered to implement their suggestions until they have been tested to see how a change in one section will affect work in other sections. Once a new idea is adopted it is documented and becomes standard practice. No deviation is allowed from a standard practice until it has been fully tested, and then if found to be a better method it will become the new standard practice. In many ways this is not far removed from Taylorism, in so far as Taylor's aim was to involve workers in finding the best method and then implementing the best method as standard practice. Over the years Taylorism became corrupted to the stage where time and motion study people (work study) found the best method and imposed the best method on the worker. With the Japanese method the workers are involved in finding the better method and of imposing the better method on themselves.

In order to alleviate the tedium of repetitive assembly work and its detrimental effect on productivity during the 1960s and 1970s many companies, especially in Scandinavia, applied two facets of job restructuring. These were job enlargement through increased task length, and job enrichment involving a greater motivational content of the job. Some companies also applied organized or self-organized job satisfaction. The case examples as reported by Wild (1975) showed improved results to a varying degree in productivity, employee turnover, absenteeism and quality. However as the emphasis of job restructuring was upon the design, or redesign, of individual manual jobs, with increased automation and mechanisation of assembly work, the impact of job restructuring became less significant. Manufacturing companies which are primarily concerned with machine and process operations rather than assembly have also sought the improvement of work organization. Thus all types of manufacturing companies are showing interest in group or team work structures. Despite the interest in team working one cannot ignore the failure of the Volvo experiment at Uddevalla where a team of workers completely assembled cars at the one static work station. The experimental factory at Uddevalla was opened in 1989 and was closed in 1992. Also closed was the Kalmar factory which had opened in 1974. (Kalmar also used a team approach to car assembly, but limited to a team completing a full element of the production process in cycles of three to four minutes.) Berggren (1994) believes that at Uddevalla, Volvo showed that there is a viable alternative to assembly line production. He lists external factors such as lack of dedicated suppliers, poor quality components, old fashioned design and so on as the reasons for failure rather than the failure of the team work approach per se. None the less, despite Berggren's well presented case, there can be little argument that the traditional assembly line model is still the most efficient method of car assembly.

In addition to the principles and case examples on group working as reported by Wild (1995), other methods of team structuring have developed more recently. One such example is Schumacher's work structuring approach (1993).

Schumacher's approach of work structure adapted tools based on industrial engineering principles, such as transformation analysis and TIED (technical, informational, error, direction) analysis. Transformation analysis involves the breaking down of tasks into basic transformation (BT), supplementary transformation (ST), ancillary operations (AO), transportation (T) and storage (S). This is comparable to the flow process chart analysis of industrial engineering comprising operation (O), inspection (◊), transport (⇨), delay (D) and storage (∇). Transformation analysis is used to streamline a task into value added (BT and ST) and cost added (AO, T and S) and also for work grouping ('whole task'). TIED analysis is designed to test the relationship between various activities by examining the links between them in each category of technical, information, error (or quality) and directional. Activities in each of these categories is scored between 0 to 4 and totals compared. The higher the score, the closer the link. TIED analysis is comparable to the critical examination of industrial engineering. Schumacher's approach of work structuring involves the following seven key principles:

1. Work should be organized around 'basic transformations' in the process of forming 'whole tasks'.
2. The basic organizational unit should be the primary work group (4 to 10 people).
3. Each work group should include a designated leader.
4. Each work group and their leader should plan and organize their own work.
5. Each work group should be able fully to evaluate its performance against agreed standards of excellence.
6. Jobs should be structured so that work group members can personally plan, do and evaluate at least one transformation in the process.
7. All work group members should have an opportunity to participate formally in the group's and the organization's common tasks.

However, as reported by Klein (1989) team structuring can also create problems. In Klein's example of 'the engine plant', team structuring included self-managed work teams with multiskilled workers, an explicit commitment to 'factory culture', achievement of schedules set by the team, and freedom to perform assembly tasks in the manner they thought best. Initially the performance of the team was highly satisfactory. But, when the company introduced a just-in-time inventory system, because the plant no longer had a buffer stock, the team came under severe delivery pressures. Just in time allowed no slack in the system and the team, without measured standards, could not meet the cost or delivery targets. In effect, the team lost its autonomy over methods and its flexibility through a lack of buffer stock. (Note the similarities to the Volvo experiment. The team was limited by the lack of support or by factors outside its direct control.)

A solution to the above problem was demonstrated by two large automobile factories. One is a GM-Toyota joint venture in California called New United Motor Manufacturing Inc., NUMMI for short (see Adler 1993) and the other is Ford-Volkswagen AG in Portugal called Autoeuropa (see Ricardo Dos Santos, 1995). In both cases the 'carte blanche' autonomy of the team was changed to a collaboration with industrial engineering. In this partnership team leaders were given comprehensive training in work study and operators (not industrial engineers) wrote methods specifications. The industrial engineer acted as a consultant or facilitator. Teams maintained their autonomy, work methods went through systematic improvement and self planning was according to measured standards. The collaboration of industrial engineering with team structuring clearly provides a balance of empowerment motivation and efficiency.

Other important factors are related to job design and team structuring ergonomics and human factors. According to a recent study by Professional Physical Therapy Services (PPTS), USA, 'The cost of implementing ergonomic changes was relatively small when compared to the total costs of functional restoration of injured persons' (*Industrial Engineering*, 1995). The common workplace injuries such as repetitive strain injuries (RSI) or operational overuse syndrome (OOS) and cumulative trauma disorder (CTD) can be prevented by appropriate workplace design and work practices. Ergonomics is a critical tool for industrial engineering.

Industrial relations

The rapid changes in flexible working practices demand an examination and reassessment of key issues in organizations and these include industrial relations practices.

During the post-war growth period trade unions in the western world assumed considerable power. The model of labour relations was basically confrontational. Relative power determined the winner of the face-to-face confrontation between management and unions. This was played out in contract negotiation and collective bargaining. Throughout the late 1960s and 1970s the recurring industrial disputes resulted in companies performing poorly. In many cases the impact was far reaching. For example the miners' strike in the UK destabilized the economy and effectively brought down the Government.

As we have indicated earlier, freer trade and deregulation along with improved logistics and communication technology now have exposed companies to global competition particularly with lower wage newly industrialized countries. The realization of this fact, particularly over the last decade, has changed the industrial relations scene. As well as structure and strength, unions have merged and have accepted the need to work with management.

At the early stage of engineering flexibility in the UK (see Cross, 1985) unions were reluctant to accept changes and signed agreements primarily because of additional 'flexibility payments'. In recent years radical changes in the business environment have prompted unions to take the initiative of formalized flexibility agreements which have helped to attract new businesses in the UK (e.g. Nissan, Siemens, Sony and Toyota).

Flexible work practices, such as the semi-autonomous or self-directed work teams, have potential problems from an industrial relations perspective. They come into conflict with job classifications and social security entitlements. Work structuring and empowerment initiatives visibly have occurred with people reduction and demands for concessions. Direct employee involvement, such as peer discipline within a team, can also undermine the role of the union. Therefore if these anomalies are not addressed with a mutual trust between management and union lasting good industrial relations will remain fragile. The success of flexible working practices will be assured only if management works with the unions rather than around the unions or against them. A formal collaboration between management and unions would eliminate some of the mutual mistrust. If a company does not have a trade union then its role should be adopted by an authoritative work council and not such token bodies whose contribution is to 'determine the colour of napkins in the canteen'.

Nancy Day reports that there are many successful models of 'union-management collaboration' (1994). One such model as promoted in Canada is called 'strategic alliance' (see Ruth Wright, 1995). This initiative contains four elements of sustainable processes:

162 TOTAL MANUFACTURING SOLUTIONS

- **Element 1: Principle-driven** Management and union must understand and accept the fundamental principle that management-union co-operation should benefit all.
- **Element 2: Fostering commitment** Commitment to the partnership can be demonstrated by contractual agreement and investment of resources.
- **Element 3: Internal and external support** Leaders of both management and unions must ensure that understanding and support are secured throughout the organization.
- **Element 4: Adherence to procedures** Procedure (or 'process') refers to the jointly determined ground rules upon which alliance stands.

Although the necessity of joint initiatives or 'strategic alliances' has been recognized, their implementations have a long way to go. Harvard Business School Professor Quin Mills (1994) found that only 20 per cent of 224 large US companies had given serious consideration to participating alliance between management and unions.

Foundation 20: Continuous learning

Human resource management can create real strategic advantage by proper planning for people with the right skills and calibre to suit the corporate strategy. If it is recognized that 'people make things happen', then there will be a continuous need for recruitment, training and development of the workforce at all levels. Schonberger (1986) describes training as the catalyst for change programmes such as JIT, TQM and BPR. Implementation plans for change management programmes usually consist of boxes with words in them connected by arrows. Such a plan for world class manufacturing comprises half a dozen arrow connected boxes, with the word training shown in each one: related to this, education and training are critical components of an empowered work environment and for an effective alliance between management and unions.

In a terminology trap, there seems to be a status distinction with the terms training and education. For example training is associated with imparting skills (how to do) and education with imparting knowledge (why it should be done). Training is thought as being needed by people at a lower level to yourself, and education is for your level and above (see also Tompkins, 1989). To overcome this implied snobbishness we use the term learning to cover both training and education. We see learning as including the development of people in their career progression.

The learning programmes of a leading-edge organization should comprise five elements:

- continuous recruitment and development,
- learning programmes for implementing new technology,
- learning for company-wide change programmes,
- learning resources, and
- learning performance.

Continuous recruitment and development

The quality of the final product depends on the materials used; therefore the development of people starts at the recruitment stage. Factories and manufacturing organizations of the future will require fewer people but the people will require a higher technical knowledge. The ability to attract high-calibre people is an indication of the sustainable performance of a company. A 'blue chip' company can attract 'high fliers'. The recruitment policy of a leading company should seek high qualifications at the entry level, for example:

- **Management trainees:** As a minimum requirement a management trainee should have a degree, with above average grades, from a reputable university.
- **Technicians:** All technicians should have a college or equivalent diploma in a suitable discipline. They will require a formal assessment after three years' training and experience before they are appointed as a technician.
- **Crafts people:** A technical high school education is the minimum requirement for a crafts person. One year of multi skill training is required before their appointment as a crafts person.
- **Operators:** Operators should have a technical high-school education.

The appointment of employees of a high educational level provides the opportunity for their development. Japanese experience in TPM clearly demonstrates that highly qualified operators can adapt, with appropriate learning opportunities, to the role of a crafts person or even a technician. A policy of employing third-party labour for operations requiring a lower skill level has been effective in many companies in the development plans of core employees.

A model for professional development programmes has been discussed in foundation stone 18, management skills and organization. It is vital that a winning company takes time to consider development plans each year so as to respond to changing needs. As Skinner has said 'If I had only one piece of advice to give to all companies, it would be move your people around' (1987).

The structure and the method of training of key operators and multiskilled crafts people should also be continuously updated. A survey by Cross (1985) indicated that of the 34 points listed for changes in systems of crafts training, the following six were seen as being necessary by most of the respondents:

(a) **Off-the-job learning benefits**
i) The development of an inquiring mind and an ability to learn for oneself.
ii) Focus in off-the-job learning is on developing motivation.

(b) **On-the-job benefits**
i) Give crafts people guidance in how to instruct.
ii) Allow supervisors to play a bigger part in giving training.
iii) Allow people to continue to receive learning as and when required over life.
iv) Reduce the risk of trainees slipping back, by providing opportunities to keep on learning.

The learning programmes for craftsmen and operators should include appraisal procedures and, where appropriate, formal lists. Key operators and crafts people should be given the opportunity to move to higher levels provided that they have met the skills, experience and performance requirements.

The effectiveness of a management appraisal programme is a key source of both the motivation and disgruntlement of managers. A poorly designed and implemented appraisal scheme can generate legions of demotivated 'under achievers'. On the other hand when the scheme contains agreed objectives based on measurable parameters and the appraiser has no axe to grind then the scheme can be effective for identifying the development needs of the appraisee. The scheme loses its credibility when the agreed action points are not followed. A leading-edge company can create a competitive edge with the aid of a properly administered appraisal scheme. The appraisal must be aimed at producing a personal development plan.

Any learning scheme should be well publicized, equitable and available to all.

Learning for implementing new technology

The learning process for implementing new technology or processes usually starts at the design stage of the project. The strategy is to grow your own in-depth skills. The example of installing a PLC-controlled high-speed laminated tube making machine for toothpaste in Chile (the first of its kind in the country) provides an example of such a training programme.

- The chief engineer, manufacturing manager and project engineer visited the supplier in Switzerland to discuss the design specifications and then visited two users (in Germany and the UK) to assess the feasibility of the project.
- After the approval of the project proposal, which included a learning and service agreement with the supplier, an operation team comprising the supervisor, technician and key operator visited the supplier for a two-week hands-on learning programme. Additional visits to operating companies who had installed the machine further motivated the team.
- The engineers from the Swiss supplier visited Chile during the commissioning of the machine. The visit plan included three days of classroom and hands-on learning. In spite of some language barriers, the engineers and key operators communicated well. The company was fortunate that they did not receive a 'sales representative' from the supplier but a genuine engineer with practical experience.
- The company formulated and installed its own internal on-the-job and off-the-job learning programme.
- The commissioning engineer (trainer) paid two more visits to Chile at six-month intervals.
- Within 18 months the operational efficiency of the machine was sustained at over 80 per cent and the in-house capability equalled that of the supplier.

Learning for a company-wide change programme

Learning is an essential component of a company-wide change programme whether it is TQM, a strategic alliance between management and unions, or a change of structure to empowered teams. Employees require learning opportunities in work process and analysis skills (e.g. SPC) as well as so called soft skills such as interrelating with team members. Line managers need to learn how to make the transition from an out-of-date autocratic management role to that of coach and mentor. The learning for all employees and managers must extend beyond skills and include learning about the need for 'cultural' change. It is essential to generate trust between all members by following the same learning process.

There are several models for company wide learning programmes. We shall mention two of them. The first one is from Oakland's 'total quality training cycle' (Oakland, 1992), as shown in Figure 8.9.

This model follows an eight step continuous process and includes a series of integrated training programmes for everyone in the organization as follows:

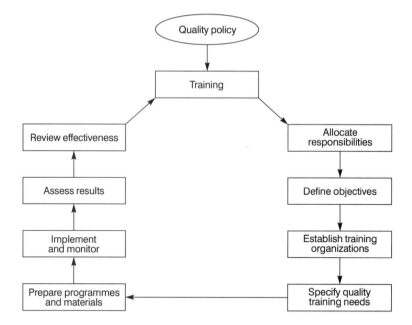

Figure 8.9 *Total quality training cycle. Source: Oakland, 1992.*

- Senior management: 8-20 hours of seminars per annum.
- Middle management: 20-30 hours seminars.
- Supervisors and team leaders: 30-40 hours.
- Seminars are followed up with workshops.
- All other employees: a half day per week for six weeks.

The second model involves the five roles put forward by Hammer and Champy (1993). These are seen to be essential to make change happen for business process re-engineering. These roles are:

1. The leader – a senior executive who authorizes and motivates the overall effort.
2. The process owner – a manager with responsibility for a specific process.
3. The re-engineering team – a group of individuals dedicated to the re-engineering of a particular process who diagnose the existing process and who oversee the redesign and the subsequent implementation.
4. The re-engineering committee – a policy making body of senior managers who develop the organization's overall re-engineering strategy and who monitor its progress.

5. The re-engineering czar – an individual responsible for developing re-engineering tools and techniques within the company and for achieving synergy across separate projects.

All the above roles need to know three things about the change process: the **why**, which is the vision and rationale for change, the **what**, which is those aspects of the organization that will be changed, and the **how**, which is the processes, tools and techniques used to design and implement the change.

Whichever model of learning programmes is applied, it is essential that management is committed and follows through with the programme, otherwise these concepts become no better than fads. With fads employees become cynical and believe that 'this too shall pass'.

Learning resources

The commitment of an organization to continuous learning is reflected in corporate support in the form of investment time, money, key personnel and facilities.

A learning organization should plan its people power to allow for the time required for both on-the-job and off-the-job learning of each employee. The total time will be variable depending on the change programmes of a particular company. For example Bell Canada allocate 15 days per year for each employee's training (see Ruth Wright, 1995).

There should be a budget for learning programmes and the measurable cost of training and education should be expressed as a percentage of annual sales. The amount allowed should be no less than 0.5 per cent of sales and in some companies it is reported to be 2 per cent of sales.

The use of a third party for specialist learning is usually successful. Each manager, supervisor and team leader should have responsibilities for the learning of their own personnel. In addition, a learning organization should provide a full-time and experienced facilitator or learning manager, to co-ordinate the continuous learning programmes of the company. For multiskilled crafts learning and TPM technical learning programmes, a competent engineering supervisor should be available to assist.

The company should provide dedicated learning facilities on site including a learning centre (equipped with personal computers, appropriate training videos and presentation facilities) and a learning workshop (equipped with work benches, mechanical, electrical and control equipment, illustrated drawings and manuals, etc.). A multi-site or multinational company should benefit from a corporate learning centre for carrying out both internal and external seminars. An on-site library of books, periodicals and internal reports is a good resource for self-education and information.

Learning performance

The effectiveness of education and training programmes of a company is usually assessed in terms of input as it is not easy to measure their output. The input measures are often expressed as:

- Number of learning hours per employee.
- Learning expenditures as a percentage of annual sales (as discussed above, up to 2 per cent of sales).
- Number of courses and seminars conducted.

Staff turnover is an indirect output measure of learning performance. A well structured management development programme, and a reputation as a company that provides good learning opportunities, will attract high-calibre candidates to an organization. A learning programme must be integrated with career development. If the management approach to human resources is well defined and considered, then most staff will repay their learning with longer service to the company. When a company loses people soon after they have been trained, then it may have got its training right but everything else is wrong.

Another assessment of learning performance involves the accreditation of learning programmes against national standards. One such scheme is 'Investor in People' in the UK. The scheme comprises four principles of employee development standards set by the Department of Education and Employment.

- A commitment from the top to develop all employees.
- Regular reviews of learning and development needs.
- Actions to train and develop individuals throughout their employment.
- Evaluation of achievement in learning and development.

The assessment indicators stem directly from the standard. If a company measures up to the national standard as assessed by Training and Enterprise Councils (TECs), then it receives an accreditation certificate of 'Investor in People'.

The drive to continuously acquire new knowledge and transfer it to skilled people to achieve its business objective is the fundamental spirit of a learning organization. A learning organization is a leading organization. The self-analysis methodology as outlined in this book presents an opportunity to identify the gaps and learning needs. An ideal map for a learning organization is shown in Figure 8.9. (See page 166.)

Summary and some final thoughts

In this chapter we have looked at people. We contend that people are the greatest resource of a company. People can be treated as servants, mere ciphers who are expected to do what they are told and to obey orders without question. Contrary to popular opinion such people do not make good soldiers. The military have long valued people with initiative and intelligence. Likewise in the factory. People power, if properly harnessed, is a tremendous force to a company seeking a competitive edge.

We began the chapter by looking at organizational structures. We hope that it would be no surprise to our readers that we favour a flat structure with a empowered work force. Such a structure requires two things, managers who are prepared to trust the workers and who are prepared to share information and decision making, and workers who are prepared to accept responsibility. To move too quickly to such a structure would be dangerous. Such a structure needs the right culture. Cultural change takes time and has to be carefully planned.

People are not fools, hidden agendas are soon exposed. People will react honestly to honesty and fair play, but once trust has gone it is hard to regain. To an extent people do only work for money, but an empowered workforce with clear guidelines and who are encouraged to make suggestions and who are listened to, will provide a higher return for the wages they receive than those who are suspicious or even resentful of the bosses. Gaining the respect of the workers means involving them in decision making, seeking their opinions and listening to them. Gone, thankfully, are the days of union and workers ranked against the bosses, of demarcation and of mutual mistrust. But we have to work hard to keep it that way.

Today we look for flexible working practices, and for unions or other representative bodies to work with management to safeguard and to grow employment opportunities. We see huge benefits accruing to a company that provides learning opportunities for all the people of the organization – opportunities which are not only on the job but off the job. We make the point that any learning plan should be open and well publicized. A set amount should be budgeted for learning each year.

In this chapter we give several models of career path development and each involves a learning component.

References

Adler, P.S. (1993) Time and Motion regained. *Harvard Business Review*, Jan-Feb, 76, No. 1.
Barlett, C. and Ghostal, S. (1994) *The Changing Role of Top Management*. INSEAD working papers, London Business School Library.

Berggren, C. (1994) *The Volvo Experience: Alternatives to Lean Production in the Swedish Auto Industry*. Macmillan, Basingstoke.
Carnall, C. (1993) *Managing Change*. Routledge, London.
Cross, M. (1985) *Towards Flexible Craftsman*. The Technical Change Centre, London.
Day, N. (1994) Human Resource Futures. *Harvard Business School Bulletin*. December.
De Meyer, A. and Ferdows, K. (1990) *Removing the Barriers in Manufacturing*. INSEAD.
Drucker, P.F. (1969) *The Age of Discontinuity*. Harper and Row.
Department of Employment, Sheffield. (1991) *Investor in People*. The National Standard.
Hamel, G. and Prahalad, E.K. (1994) Competing for the Future. *Harvard Business Review*, Jul-Aug, 72, No. 4.
Hammer, M. and Champy, J. (1993) *Reengineering the Corporation*. Harper Business.
Hay Group (1995) *People and Performance*. (Spring issue) Hay Group, London.
Professional Physical Therapy Services (1995) *Industrial Engineering*, April, p. 11.
Klein, J.A. (1989) The Human Cost of Manufacturing Reform. *Harvard Business Review*, March-April, 67, No. 2.
Kolodny, H. (1993) *The Organization Change Process*. University of Toronto.
Oakland, J.S. (1992) *Total Quality Management*. Butterworth-Heinemann, Oxford.
Ricardo Dos Santos, J. (1995) Re-Tayloring the Shopfloor. *Industrial Engineering*. May.
Schonberger, R. (1986) *World Class Manufacturing*. Free Press.
Schumacher, C. (1993) Seminar Notes on Workstructuring. Workstructuring Ltd, Godstone, Surrey.
Skinner, W. (1987) Wanted: A New Breed of Manufacturing Manager. *Manufacturing Issues*, Booz Allen and Hamilton.
The Economist (1995) The Changing Nature of Leadership. June.
Tompkins, J.A. (1989) *Winning Manufacturing*. Engineering and Management Press.
Wild, R. (1975) *Work Organization*. John Wiley & Sons.
Wright, R. (1995) *Managing Labour Relations in a New Economy*. The Conference Branch of Canada.

9

200 questions

> *I keep six honest working men*
> *(They taught me all I knew)*
> *Their names are What and Why and When*
> *and How and Where and Who.*
> Rudyard Kipling, The Elephant's Child

The 200 questions listed in this chapter constitute a central part of our approach to total manufacturing solutions. Hence, the questions are in a chapter and not in an appendix.

The 200 questions are based upon our analysis and explanation of the 20 foundation stones of total manufacturing as described in Chapters 3 to 8. The rationale for the questions is given in Chapter 2. Our emphasis is on a 'self-assessment' approach by a company's own managers and staff. In the following chapters we will discuss how a multi-functional team within a company can carry out its own benchmarking to identify areas for progressive improvement. We also discuss in subsequent chapters the role of external consultants in this exercise.

As explained in Chapter 1, we have 10 questions for each of the 20 foundation stones. We have listed and numbered the questions in the same order as we introduced each foundation stone. The sequence of the questions does not imply any relative importance.

The questions have been constructed so that they are easy to interpret. If there are any difficulties refer back to the relevant pillar/foundation stone chapter. The scoring by an experienced team for each question; 0.1 (poor), 0.2 (fair), 0.3 (good), 0.4 (very good) and 0.5 (excellent), permits a high degree of accuracy. We discuss guidelines for team members in Chapter 10.

When we use the word you with questions, it can be taken to refer to you the reader, you as a member of a multi functional bench marking team, or to you in the sense of your company (and your company's management and planning team).

Now for the questions!

1. Understanding the marketplace

Tick appropriate box

1. How well do your managers in marketing and sales know the importance of main products (by volume, profit and trends)?
   ☐ 0.1 ☐ 0.2 ☐ 0.3 ☐ 0.4 ☐ 0.5

2. How good (precise) are your analyses of trade needs and consumer habits? (If you have your own consumer studies centre and it is effective a 0.5 rating is likely.)
   ☐ 0.1 ☐ 0.2 ☐ 0.3 ☐ 0.4 ☐ 0.5

3. How often do you conduct market research of trade needs and consumer habits? (Guess work = 0.1, but if at least every year then a rating of 0.5 is warranted.)
   ☐ 0.1 ☐ 0.2 ☐ 0.3 ☐ 0.4 ☐ 0.5

4. How often do you evaluate product performance in the market? (Less than every two years = 0.1, at least every six months = 0.5.)
   ☐ 0.1 ☐ 0.2 ☐ 0.3 ☐ 0.4 ☐ 0.5

5. How well do your sales and marketing team know the relative importance of factors that affect customer satisfaction? (Factors to consider are cost, quality, lead time, order fill, after-sales service.)
   ☐ 0.1 ☐ 0.2 ☐ 0.3 ☐ 0.4 ☐ 0.5

6. How systematic, and scientific, are your advertising and promotion activities?
   ☐ 0.1 ☐ 0.2 ☐ 0.3 ☐ 0.4 ☐ 0.5

7. How close is the link between your sales, marketing, planning and manufacturing functions?
   ☐ 0.1 ☐ 0.2 ☐ 0.3 ☐ 0.4 ☐ 0.5

8. How good is your manufacturing manager's first hand knowledge of the marketplace?
   ☐ 0.1 ☐ 0.2 ☐ 0.3 ☐ 0.4 ☐ 0.5

9. How well do you know international tariff, tax and trade regulations?
   ☐ 0.1 ☐ 0.2 ☐ 0.3 ☐ 0.4 ☐ 0.5

10. How aware are you of opportunities and constraints for emerging markets (e.g. East Europe and China)?
    ☐ 0.1 ☐ 0.2 ☐ 0.3 ☐ 0.4 ☐ 0.5

2. Understanding the competition

11. How well do you know the true market size and market share for your core products? ☐ 0.1 ☐ 0.2 ☐ 0.3 ☐ 0.4 ☐ 0.5
12. How well do your managers in marketing, sales and manufacturing know (and agree) who your top three competitors are in specific product categories? ☐ 0.1 ☐ 0.2 ☐ 0.3 ☐ 0.4 ☐ 0.5
13. How well and how frequently is product by growth, market share, profitability, etc., analysed (for example BCG matrix)? ☐ 0.1 ☐ 0.2 ☐ 0.3 ☐ 0.4 ☐ 0.5
14. How good is your knowledge of the strengths and weaknesses of your top three competitors? ☐ 0.1 ☐ 0.2 ☐ 0.3 ☐ 0.4 ☐ 0.5
15. How well do you know and compare the service level which your key competitors provide to your customers? ☐ 0.1 ☐ 0.2 ☐ 0.3 ☐ 0.4 ☐ 0.5
16. How well do you know your competitors' innovation programmes and success rates? ☐ 0.1 ☐ 0.2 ☐ 0.3 ☐ 0.4 ☐ 0.5
17. How well do you know your key competitors' acquisitions, expansion and divestment programmes? ☐ 0.1 ☐ 0.2 ☐ 0.3 ☐ 0.4 ☐ 0.5
18. How well do you know the capacities and utilization of competitors' manufacturing units and distribution centres? ☐ 0.1 ☐ 0.2 ☐ 0.3 ☐ 0.4 ☐ 0.5
19. How effectively do you keep track of the emergence of new competitors? ☐ 0.1 ☐ 0.2 ☐ 0.3 ☐ 0.4 ☐ 0.5
20. How often do you take part in external benchmarking surveys related to your competition? (Seldom = 0.1, once a year = 0.5.) ☐ 0.1 ☐ 0.2 ☐ 0.3 ☐ 0.4 ☐ 0.5

3. Product and process innovation

21. How generous is your budget for product and process innovation? (<0.25% of sales = 0.1, >2% of sales = 0.5) ☐ 0.1 ☐ 0.2 ☐ 0.3 ☐ 0.4 ☐ 0.5
22. How well has your product innovation lead time reduced over the last three years? ☐ 0.1 ☐ 0.2 ☐ 0.3 ☐ 0.4 ☐ 0.5
23. How good is your success ratio for new products? (<10% = 0.1, >80% = 0.5) ☐ 0.1 ☐ 0.2 ☐ 0.3 ☐ 0.4 ☐ 0.5

24. How well is product design carried out in parallel with process design?
    ☐ 0.1 ☐ 0.2 ☐ 0.3 ☐ 0.4 ☐ 0.5
25. How well do you use an integrated multidisciplined team for product development with *full* participation of manufacturing?
    ☐ 0.1 ☐ 0.2 ☐ 0.3 ☐ 0.4 ☐ 0.5
26. How often and how well do you design new products with a focus on ease of manufacturing?
    ☐ 0.1 ☐ 0.2 ☐ 0.3 ☐ 0.4 ☐ 0.5
27. How well do you apply project management in innovation?
    ☐ 0.1 ☐ 0.2 ☐ 0.3 ☐ 0.4 ☐ 0.5
28. How effectively do you structure and apply a screening process so as to eliminate impractical ideas at each stage of development?
    ☐ 0.1 ☐ 0.2 ☐ 0.3 ☐ 0.4 ☐ 0.5
29. How well do you carry out the analysis of product life cycles? (Not very well = 0.1, continuous = 0.5.)
    ☐ 0.1 ☐ 0.2 ☐ 0.3 ☐ 0.4 ☐ 0.5
30. How good and how frequent is your evaluation of your manufacturing capability to meet the impact of innovation?
    ☐ 0.1 ☐ 0.2 ☐ 0.3 ☐ 0.4 ☐ 0.5

4. Manufacturing resources planning and working with suppliers

31. How well is a business planning process used to develop annual financial plans?
    ☐ 0.1 ☐ 0.2 ☐ 0.3 ☐ 0.4 ☐ 0.5
32. How well do your sales and operations planning support the business plan? (Irregular = 0.1, very effective with weekly reviews = 0.5.)
    ☐ 0.1 ☐ 0.2 ☐ 0.3 ☐ 0.4 ☐ 0.5
33. How good is your forecasting of anticipated demand to support planning? (Consider attention to promotion, seasonality and marketing intelligence.)
    ☐ 0.1 ☐ 0.2 ☐ 0.3 ☐ 0.4 ☐ 0.5
34. How effective is your master scheduling process to ensure a balance between the sales plan and the stock of finished products?
    ☐ 0.1 ☐ 0.2 ☐ 0.3 ☐ 0.4 ☐ 0.5
35. How good is your capacity planning? Do you use a rough cut capacity plan to develop a detailed capacity requirement plan?
    ☐ 0.1 ☐ 0.2 ☐ 0.3 ☐ 0.4 ☐ 0.5
36. How well is your purchase scheduling managed? What controls are there in place?
    ☐ 0.1 ☐ 0.2 ☐ 0.3 ☐ 0.4 ☐ 0.5

37. How well do you pursue a make-to-order policy with an emphasis on material velocity (stock turn)? (Large stocks of raw materials, work in progress and finished goods = 0.1, just-in-time philosophy with little or no buffer stock = 0.5.)
☐ 0.1 ☐ 0.2 ☐ 0.3 ☐ 0.4 ☐ 0.5

38. How effective is your long-term partnership with key suppliers? (Large number of competing suppliers = 0.1. Key suppliers included as part of the planning team = 0.5.)
☐ 0.1 ☐ 0.2 ☐ 0.3 ☐ 0.4 ☐ 0.5

39. How successful have you been with the co-development of new materials?
☐ 0.1 ☐ 0.2 ☐ 0.3 ☐ 0.4 ☐ 0.5

40. How effective have you been in the sharing of common coding and databases facilitated by electronic data interchange (EDI)?
☐ 0.1 ☐ 0.2 ☐ 0.3 ☐ 0.4 ☐ 0.5

5. Distribution management and working with customers

41. How well is your distribution strategy defined and is it regularly reviewed? Does your strategy include third-party warehousing/transport?
☐ 0.1 ☐ 0.2 ☐ 0.3 ☐ 0.4 ☐ 0.5

42. How good is your design (type, size, location, etc.) and the operation of your warehouses (own and third party)?
☐ 0.1 ☐ 0.2 ☐ 0.3 ☐ 0.4 ☐ 0.5

43. How often do you experience stock outs of raw materials and/or finished products?
☐ 0.1 ☐ 0.2 ☐ 0.3 ☐ 0.4 ☐ 0.5

44. How effective is your transport planning for both primary (trunking) and secondary (local delivery) operations?
☐ 0.1 ☐ 0.2 ☐ 0.3 ☐ 0.4 ☐ 0.5

45. How good is your formal distribution resource planning for each stock keeping unit (SKU)?
☐ 0.1 ☐ 0.2 ☐ 0.3 ☐ 0.4 ☐ 0.5

46. How effectively does your organization span (across functions) the whole supply chain? (One executive responsible for planning, production and distribution = 0.5.)
☐ 0.1 ☐ 0.2 ☐ 0.3 ☐ 0.4 ☐ 0.5

47. How closely is your manufacturing manager involved with customers to achieve a precise specification of customers' needs?
☐ 0.1 ☐ 0.2 ☐ 0.3 ☐ 0.4 ☐ 0.5

48. How frequently and effectively do you analyse channels of distribution (e.g. supermarket, wholesaler, retailer)?
    ☐ 0.1 ☐ 0.2 ☐ 0.3 ☐ 0.4 ☐ 0.5
49. How well do you measure customer profitability by activity-based cost location?
    ☐ 0.1 ☐ 0.2 ☐ 0.3 ☐ 0.4 ☐ 0.5
50. Have you established a serious partnership with key customers to help them gain a competitive edge by providing an all round package of customer service?
    ☐ 0.1 ☐ 0.2 ☐ 0.3 ☐ 0.4 ☐ 0.5

6. Supply-chain performance

51. How good is your *stock turn* and how well does it match your stock policy?
    ☐ 0.1 ☐ 0.2 ☐ 0.3 ☐ 0.4 ☐ 0.5
52. How good is your *planning efficiency* by product group?
    ☐ 0.1 ☐ 0.2 ☐ 0.3 ☐ 0.4 ☐ 0.5
53. How good is your *asset turn* expressed as a ratio of sales and fixed assets?
    ☐ 0.1 ☐ 0.2 ☐ 0.3 ☐ 0.4 ☐ 0.5
54. How good are you in meeting *on-time delivery* as determined by the standards of order cycle time set for customers?
    ☐ 0.1 ☐ 0.2 ☐ 0.3 ☐ 0.4 ☐ 0.5
55. How well do you perform in satisfying orders at brand and SKU level?
    ☐ 0.1 ☐ 0.2 ☐ 0.3 ☐ 0.4 ☐ 0.5
56. How satisfactory is your *post-delivery performance* in terms of invoice accuracy?
    ☐ 0.1 ☐ 0.2 ☐ 0.3 ☐ 0.4 ☐ 0.5
57. How satisfactory is your quality performance measured in terms of *return of goods* at brand and SKU level?
    ☐ 0.1 ☐ 0.2 ☐ 0.3 ☐ 0.4 ☐ 0.5
58. How good is your *composite customer service* performance by weighting the key supply chain performance indicators?
    ☐ 0.1 ☐ 0.2 ☐ 0.3 ☐ 0.4 ☐ 0.5
59. How cost efficient is your distribution operation when distribution cost is expressed as a percentage of sales? (Over 8% = 0.1, excellent = less than 1% = 0.5.)
    ☐ 0.1 ☐ 0.2 ☐ 0.3 ☐ 0.4 ☐ 0.5
60. How easy is it for customers to contact the right person in your organization when they want to place an order or need knowledge of a product?
    ☐ 0.1 ☐ 0.2 ☐ 0.3 ☐ 0.4 ☐ 0.5

7. Product safety

61. How effective is your safety clearance of new products particularly regarding toxicological hazards? (100% clearance mandatory = 0.5.) ☐ 0.1 ☐ 0.2 ☐ 0.3 ☐ 0.4 ☐ 0.5
62. How well do you maintain product formulation and packaging standards (e.g. to avoid over packaging)? ☐ 0.1 ☐ 0.2 ☐ 0.3 ☐ 0.4 ☐ 0.5
63. How well is the receipt of raw and packaging materials controlled? Should there be laboratory tests as part of your receiving process (e.g. microbiological tests)? (100% for hazard materials, monitor approved suppliers with sampling, etc. = 0.5.) ☐ 0.1 ☐ 0.2 ☐ 0.3 ☐ 0.4 ☐ 0.5
64. How good is your installation and monitoring of water decontamination systems (chlorination, pasteurization)? ☐ 0.1 ☐ 0.2 ☐ 0.3 ☐ 0.4 ☐ 0.5
65. How well have you incorporated hygiene design aspects in manufacturing plants, especially for food products (cleaning in place (CIP) procedures)? ☐ 0.1 ☐ 0.2 ☐ 0.3 ☐ 0.4 ☐ 0.5
66. How good are your quality assurance procedures and control at each stage of the supply chain including sanitation, house keeping, storage and handling? ☐ 0.1 ☐ 0.2 ☐ 0.3 ☐ 0.4 ☐ 0.5
67. How good are your selection and use of preservatives to prevent accidental contamination? ☐ 0.1 ☐ 0.2 ☐ 0.3 ☐ 0.4 ☐ 0.5
68. How good are your procedures for handling consumer complaints and product safety incidents including product recall procedures? ☐ 0.1 ☐ 0.2 ☐ 0.3 ☐ 0.4 ☐ 0.5
69. How good are your on-site laboratory facilities and resources? (Including a QA manager qualified in microbiology where appropriate.) ☐ 0.1 ☐ 0.2 ☐ 0.3 ☐ 0.4 ☐ 0.5
70. How good and effective are your learning programmes for all employees regarding product safety? ☐ 0.1 ☐ 0.2 ☐ 0.3 ☐ 0.4 ☐ 0.5

8. Industrial safety

71. How effective is your HAZOP system before and after the installation of a plant? (Both before and after, = 0.5.) ☐ 0.1 ☐ 0.2 ☐ 0.3 ☐ 0.4 ☐ 0.5

72. How effective are your accident prevention system and maintenance of equipment (e.g. low-voltage equipment, fire walls, machine guards, sprinklers, smoke alarms, etc.)? ☐ 0.1 ☐ 0.2 ☐ 0.3 ☐ 0.4 ☐ 0.5

73. How effective are your emergency procedures? (Fire fighting equipment, evacuation routes and procedures. Are drills carried out regularly? etc.) ☐ 0.1 ☐ 0.2 ☐ 0.3 ☐ 0.4 ☐ 0.5

74. How good is your engineering control of hazardous material and equipment? (Current permits, scheduled maintenance, regular inspections, dangerous goods stores, etc.) ☐ 0.1 ☐ 0.2 ☐ 0.3 ☐ 0.4 ☐ 0.5

75. How good is your control and monitoring of exposure to dust, fumes, hazardous substances, noise? ☐ 0.1 ☐ 0.2 ☐ 0.3 ☐ 0.4 ☐ 0.5

76. How well have you implemented personal safety and protection measures, especially for employees in hazardous areas? ☐ 0.1 ☐ 0.2 ☐ 0.3 ☐ 0.4 ☐ 0.5

77. How good is your organization of resources for first aid, professional nursing care, health and safety stewards, fire wardens, etc.? ☐ 0.1 ☐ 0.2 ☐ 0.3 ☐ 0.4 ☐ 0.5

78. How well have you applied an accredited safety audit (e.g. ISRS Certificate)? ☐ 0.1 ☐ 0.2 ☐ 0.3 ☐ 0.4 ☐ 0.5

79. How effective are your learning programmes for all employees regarding industrial safety? ☐ 0.1 ☐ 0.2 ☐ 0.3 ☐ 0.4 ☐ 0.5

80. How well do you monitor the statistics of losses due to accidents? e.g. accident rate (accident incidents x 2000 divided by employee hours.) ☐ 0.1 ☐ 0.2 ☐ 0.3 ☐ 0.4 ☐ 0.5

9. Environmental protection

81. How good are your plant and control systems regarding the treatment of aqueous effluent, gaseous emission and solid waste disposal? ☐ 0.1 ☐ 0.2 ☐ 0.3 ☐ 0.4 ☐ 0.5

82. How good are your facilities and systems for monitoring and controlling noise to an acceptable level? ☐ 0.1 ☐ 0.2 ☐ 0.3 ☐ 0.4 ☐ 0.5
83. How well are you complying with legislation and measuring up to public concerns regarding environmental issues? ☐ 0.1 ☐ 0.2 ☐ 0.3 ☐ 0.4 ☐ 0.5
84. How effective are you in avoiding/detecting pollutant metals and materials in plant and production (such as lead, cadmium, mercury, and asbestos)? ☐ 0.1 ☐ 0.2 ☐ 0.3 ☐ 0.4 ☐ 0.5
85. How good are you in your energy policy with a focus on environmental issues? Does a qualified person have responsibility for keeping your policy current? ☐ 0.1 ☐ 0.2 ☐ 0.3 ☐ 0.4 ☐ 0.5
86. How effective are you in 'non-waste technology' (e.g. converting suspended solids into animal feeds)? ☐ 0.1 ☐ 0.2 ☐ 0.3 ☐ 0.4 ☐ 0.5
87. How well do you negotiate emission standards with government authorities? ☐ 0.1 ☐ 0.2 ☐ 0.3 ☐ 0.4 ☐ 0.5
88. How good are your emergency procedures for the accidental release of pollutants? ☐ 0.1 ☐ 0.2 ☐ 0.3 ☐ 0.4 ☐ 0.5
89. How effective are your environmental awareness learning sessions for all your personnel? ☐ 0.1 ☐ 0.2 ☐ 0.3 ☐ 0.4 ☐ 0.5
90. How well do you monitor the safety and environmental standards of your business partners (e.g. raw material suppliers, co-packers, co-producers, contractors)? ☐ 0.1 ☐ 0.2 ☐ 0.3 ☐ 0.4 ☐ 0.5

10. Sourcing strategy

91. How forward looking is your written sourcing and manufacturing strategy? ☐ 0.1 ☐ 0.2 ☐ 0.3 ☐ 0.4 ☐ 0.5
92. How relevant (effective) is your written manufacturing mission? ☐ 0.1 ☐ 0.2 ☐ 0.3 ☐ 0.4 ☐ 0.5
93. How well quantified are your manufacturing objectives (are outcomes measurable)? ☐ 0.1 ☐ 0.2 ☐ 0.3 ☐ 0.4 ☐ 0.5
94. How well have you developed master plans for your manufacturing sites? ☐ 0.1 ☐ 0.2 ☐ 0.3 ☐ 0.4 ☐ 0.5
95. How good and reliable is your long-term forecast of sales volume by main product groups? ☐ 0.1 ☐ 0.2 ☐ 0.3 ☐ 0.4 ☐ 0.5

96. How good and reliable are your analyses of plant capacity and cost information? ☐ 0.1 ☐ 0.2 ☐ 0.3 ☐ 0.4 ☐ 0.5
97. How good is your evaluation of strategic options and servicing options? (Subjective = 0.1, Simulation modelling and SWOT analysis, = 0.5.) ☐ 0.1 ☐ 0.2 ☐ 0.3 ☐ 0.4 ☐ 0.5
98. How well have you analysed your competitors' game plans and strengths? ☐ 0.1 ☐ 0.2 ☐ 0.3 ☐ 0.4 ☐ 0.5
99. How forward looking is your capital expenditure plan? ☐ 0.1 ☐ 0.2 ☐ 0.3 ☐ 0.4 ☐ 0.5
100. How well and how often do you review your sourcing and manufacturing strategy with regard to changing conditions? ☐ 0.1 ☐ 0.2 ☐ 0.3 ☐ 0.4 ☐ 0.5

11. Appropriate manufacturing technology

101. How well are volume and growth considered when selecting technology? ☐ 0.1 ☐ 0.2 ☐ 0.3 ☐ 0.4 ☐ 0.5
102. How well are product variety (and trends) considered when selecting technology? ☐ 0.1 ☐ 0.2 ☐ 0.3 ☐ 0.4 ☐ 0.5
103. How well do you keep abreast of best practice in use and changes in technology? ☐ 0.1 ☐ 0.2 ☐ 0.3 ☐ 0.4 ☐ 0.5
104. How well do you consider your operators' experience and skills when selecting technology? ☐ 0.1 ☐ 0.2 ☐ 0.3 ☐ 0.4 ☐ 0.5
105. How well is advantage taken of the capacity, experience and local representation of suppliers of new technology? ☐ 0.1 ☐ 0.2 ☐ 0.3 ☐ 0.4 ☐ 0.5
106. In the analysis of new technology proposals, how effectively are alternative solutions and value engineering used or considered? ☐ 0.1 ☐ 0.2 ☐ 0.3 ☐ 0.4 ☐ 0.5
107. How good is your evaluation of investment proposals for new technology in terms of return on investment? Conversely, is there a danger that ROI analysis will leave you with out dated equipment? ☐ 0.1 ☐ 0.2 ☐ 0.3 ☐ 0.4 ☐ 0.5
108. How good is your strategic approach to new technology (e.g. are you aware of new technology being introduced by competitors)? ☐ 0.1 ☐ 0.2 ☐ 0.3 ☐ 0.4 ☐ 0.5

109. When evaluating new technology how well do you use the experience of other users? How vulnerable is the new technology?

    ☐ 0.1 ☐ 0.2 ☐ 0.3 ☐ 0.4 ☐ 0.5

110. How good is your technology development plan (both product and process) with regard to your resources and needs?

    ☐ 0.1 ☐ 0.2 ☐ 0.3 ☐ 0.4 ☐ 0.5

12. Flexible manufacturing systems

111. How good is your understanding of flexibility in manufacturing?

    ☐ 0.1 ☐ 0.2 ☐ 0.3 ☐ 0.4 ☐ 0.5

112. How well do your managers appreciate the need for flexibility?

    ☐ 0.1 ☐ 0.2 ☐ 0.3 ☐ 0.4 ☐ 0.5

113. How closely do marketing and manufacturing work together to standardize common elements so as to minimize the need for flexibility?

    ☐ 0.1 ☐ 0.2 ☐ 0.3 ☐ 0.4 ☐ 0.5

114. How quickly can you respond to changes in volume and variety by flexible work systems?

    ☐ 0.1 ☐ 0.2 ☐ 0.3 ☐ 0.4 ☐ 0.5

115. Without affecting the market, how effective is your product and material harmonization?

    ☐ 0.1 ☐ 0.2 ☐ 0.3 ☐ 0.4 ☐ 0.5

116. How flexible is your manufacturing plant/equipment?

    ☐ 0.1 ☐ 0.2 ☐ 0.3 ☐ 0.4 ☐ 0.5

117. How well have you achieved rapid and minimum change-overs in production (e.g. single minute exchange of dies, SMED)?

    ☐ 0.1 ☐ 0.2 ☐ 0.3 ☐ 0.4 ☐ 0.5

118. How well do you cope with changes to production schedules (is planning data sufficiently reliable to permit rapid changes)?

    ☐ 0.1 ☐ 0.2 ☐ 0.3 ☐ 0.4 ☐ 0.5

119. How effective have you been in the modular design of equipment and facilities?

    ☐ 0.1 ☐ 0.2 ☐ 0.3 ☐ 0.4 ☐ 0.5

120. Do you have effective alternative finishing stages (e.g. packing line options)?

    ☐ 0.1 ☐ 0.2 ☐ 0.3 ☐ 0.4 ☐ 0.5

13. Reliable manufacturing

121. How well do your manufacturing managers understand the ramifications of the visible costs of maintenance and less tangible maintenance failure related costs? ☐ 0.1 ☐ 0.2 ☐ 0.3 ☐ 0.4 ☐ 0.5
122. How good are your written maintenance policies for different types of assets? ☐ 0.1 ☐ 0.2 ☐ 0.3 ☐ 0.4 ☐ 0.5
123. How well do you focus on right-first-time reliable equipment for new investments and continuously seek to improve/modify existing equipment? ☐ 0.1 ☐ 0.2 ☐ 0.3 ☐ 0.4 ☐ 0.5
124. How well do you follow predictive and condition-based maintenance principles? ☐ 0.1 ☐ 0.2 ☐ 0.3 ☐ 0.4 ☐ 0.5
125. How good are your maintenance staff in terms of specialist skills and experience? ☐ 0.1 ☐ 0.2 ☐ 0.3 ☐ 0.4 ☐ 0.5
126. How good is your maintenance support infrastructure in terms of own workshops, technical stores and third-party capability? ☐ 0.1 ☐ 0.2 ☐ 0.3 ☐ 0.4 ☐ 0.5
127. How effective are your maintenance information and management systems for the control of assets, spares, planning and cost? ☐ 0.1 ☐ 0.2 ☐ 0.3 ☐ 0.4 ☐ 0.5
128. How effectively have you achieved autonomous maintenance (empowered work force) with TPM? ☐ 0.1 ☐ 0.2 ☐ 0.3 ☐ 0.4 ☐ 0.5
129. How well is it understood throughout your organization that maintenance is a competitive weapon? ☐ 0.1 ☐ 0.2 ☐ 0.3 ☐ 0.4 ☐ 0.5
130. How good are your maintenance effectiveness ratios (e.g. as % CRV, % sales, % factory operating costs)? ☐ 0.1 ☐ 0.2 ☐ 0.3 ☐ 0.4 ☐ 0.5

14. Manufacturing performance

131. How good are your manufacturing performance measurements and how well are they monitored? ☐ 0.1 ☐ 0.2 ☐ 0.3 ☐ 0.4 ☐ 0.5
132. How good are your labour productivity and trend? ☐ 0.1 ☐ 0.2 ☐ 0.3 ☐ 0.4 ☐ 0.5
133. How good are your materials productivity and trend? ☐ 0.1 ☐ 0.2 ☐ 0.3 ☐ 0.4 ☐ 0.5

134. How good are your energy economy and trend? ☐ 0.1 ☐ 0.2 ☐ 0.3 ☐ 0.4 ☐ 0.5
135. How good are your plant efficiency and trend? (< 45% = 0.1, > 80% = 0.5) ☐ 0.1 ☐ 0.2 ☐ 0.3 ☐ 0.4 ☐ 0.5
136. How good are your plant utilization and trend? ☐ 0.1 ☐ 0.2 ☐ 0.3 ☐ 0.4 ☐ 0.5
137. How good are your internal effectiveness figures (e.g. quality, maintenance, safety) and trends? ☐ 0.1 ☐ 0.2 ☐ 0.3 ☐ 0.4 ☐ 0.5
138. How good are your external effectiveness figures (e.g. innovation, customer service) and trends? ☐ 0.1 ☐ 0.2 ☐ 0.3 ☐ 0.4 ☐ 0.5
139. How effective and team based are your performance improvement programmes (e.g. TPM)? ☐ 0.1 ☐ 0.2 ☐ 0.3 ☐ 0.4 ☐ 0.5
140. How good are your resources (e.g. trained industrial engineers) for monitoring and improving performance? ☐ 0.1 ☐ 0.2 ☐ 0.3 ☐ 0.4 ☐ 0.5

15. Quality management

141. How well practised is the total quality management concept in your organization, including customer supplier relationships internally and externally? ☐ 0.1 ☐ 0.2 ☐ 0.3 ☐ 0.4 ☐ 0.5
142. How effective are the commitments and support of top management in quality management? ☐ 0.1 ☐ 0.2 ☐ 0.3 ☐ 0.4 ☐ 0.5
143. How well do you quantify quality targets and performance? (Are results displayed at the work place?) ☐ 0.1 ☐ 0.2 ☐ 0.3 ☐ 0.4 ☐ 0.5
144. How effectively do you focus on preventing the cost of non-conformance? ☐ 0.1 ☐ 0.2 ☐ 0.3 ☐ 0.4 ☐ 0.5
145. How well is the use of SPC tools understood at all levels? ☐ 0.1 ☐ 0.2 ☐ 0.3 ☐ 0.4 ☐ 0.5
146. How well do you monitor the quality standards of approved suppliers? ☐ 0.1 ☐ 0.2 ☐ 0.3 ☐ 0.4 ☐ 0.5
147. How clear is your written policy on quality? Is it readily available throughout your organization? ☐ 0.1 ☐ 0.2 ☐ 0.3 ☐ 0.4 ☐ 0.5
148. How well documented is your standard procedure for all processes? Are your procedures reviewed and updated regularly? ☐ 0.1 ☐ 0.2 ☐ 0.3 ☐ 0.4 ☐ 0.5

184 TOTAL MANUFACTURING SOLUTIONS

149. How successful have you been in achieving a quality award or accreditation to a quality standard (e.g. a Baldridge award or ISO 9000 accreditation)? ☐ 0.1 ☐ 0.2 ☐ 0.3 ☐ 0.4 ☐ 0.5

150. How effective are your learning programmes for TQM at all levels? ☐ 0.1 ☐ 0.2 ☐ 0.3 ☐ 0.4 ☐ 0.5

16. Financial management

151. How well and clearly have key financial parameters for manufacturing been written into corporate strategic and annual operating plans? ☐ 0.1 ☐ 0.2 ☐ 0.3 ☐ 0.4 ☐ 0.5

152. How well do your manufacturing managers understand financial parameters and cost structures? ☐ 0.1 ☐ 0.2 ☐ 0.3 ☐ 0.4 ☐ 0.5

153. What is your profit margin as a percentage of sales (return on sales? Is the trend upwards?)?
(Less than 5% = 0.1, over 15% = 0.5.) ☐ 0.1 ☐ 0.2 ☐ 0.3 ☐ 0.4 ☐ 0.5

154. How sound is your working capital position (have assets been sold to improve the position or was improvement from operating results)? ☐ 0.1 ☐ 0.2 ☐ 0.3 ☐ 0.4 ☐ 0.5

155. How good is your return on capital employed (ROI)? (Less than 6% = 0.1, over 20% = 0.5.) ☐ 0.1 ☐ 0.2 ☐ 0.3 ☐ 0.4 ☐ 0.5

156. How good is your cash flow (consider debtors turn and stock turn ratios)? ☐ 0.1 ☐ 0.2 ☐ 0.3 ☐ 0.4 ☐ 0.5

157. How rigorous and effective is your appraisal of capital proposals? ☐ 0.1 ☐ 0.2 ☐ 0.3 ☐ 0.4 ☐ 0.5

158. How effective and up to date is your accounting system? Is information relevant, clearly presented, timely and accurate? ☐ 0.1 ☐ 0.2 ☐ 0.3 ☐ 0.4 ☐ 0.5

159. How good are your cost-effectiveness programmes and achievements? ☐ 0.1 ☐ 0.2 ☐ 0.3 ☐ 0.4 ☐ 0.5

160. How good is the trend of your share price on the stock market? ☐ 0.1 ☐ 0.2 ☐ 0.3 ☐ 0.4 ☐ 0.5

17. Information technology

161. How good is your written IT strategy for the hardware platform and operating systems regarding open systems (e.g. OSF/AES) standards? 0.1 0.2 0.3 0.4 0.5
162. How good are your hardware support agreements? 0.1 0.2 0.3 0.4 0.5
163. How effective is your software strategy regarding selection of packages and developing own systems? 0.1 0.2 0.3 0.4 0.5
164. How good is the application software for commercial and financial management? 0.1 0.2 0.3 0.4 0.5
165. How good are the software packages for design, factory automation and factory administration? 0.1 0.2 0.3 0.4 0.5
166. How effective is the software for supply chain management including MRP II? 0.1 0.2 0.3 0.4 0.5
167. How effective is integration of data and application software regarding data architecture, local area network and wide area network? 0.1 0.2 0.3 0.4 0.5
168. How effective are the 'client' PC work stations regarding software (word processing, Windows, Lotus, conference, E-mail, etc.) and networking? 0.1 0.2 0.3 0.4 0.5
169. How good are your own IT resources and organization at developing and supporting IT applications? Is there a disaster recovery plan? Has it been tested? 0.1 0.2 0.3 0.4 0.5
170. How effective is your strategy for selection, prototyping, training and implementing a company-wide IT application? 0.1 0.2 0.3 0.4 0.5

18. Management skills and organization

171. How successful and effective have you been in implementing a 'flatter' organization with a reduced number of 'layers' of managers? 0.1 0.2 0.3 0.4 0.5
172. How effective is your organization in integrating supply-chain related activities under one senior manager? 0.1 0.2 0.3 0.4 0.5

173. How good is your policy to attract, develop and retain high-calibre managers? ☐ 0.1 ☐ 0.2 ☐ 0.3 ☐ 0.4 ☐ 0.5
174. How good are your professional staff in commercial, financial and human resources management? ☐ 0.1 ☐ 0.2 ☐ 0.3 ☐ 0.4 ☐ 0.5
175. How good and motivated is your marketing and sales force? ☐ 0.1 ☐ 0.2 ☐ 0.3 ☐ 0.4 ☐ 0.5
176. How good are your technical managers with regard to their knowledge of technology in their area of expertise and do they have a good understanding of other functions? ☐ 0.1 ☐ 0.2 ☐ 0.3 ☐ 0.4 ☐ 0.5
177. How effective is your policy for transferring people between operations and strategic 'think tanks'? ☐ 0.1 ☐ 0.2 ☐ 0.3 ☐ 0.4 ☐ 0.5
178. How good is your projection of corporate culture and identity through a set of written strategies? ☐ 0.1 ☐ 0.2 ☐ 0.3 ☐ 0.4 ☐ 0.5
179. How effective is your leadership as demonstrated by a vision of competing for the future? ☐ 0.1 ☐ 0.2 ☐ 0.3 ☐ 0.4 ☐ 0.5
180. How effective is your leadership as demonstrated by encouraging entrepreneurism of smaller units? ☐ 0.1 ☐ 0.2 ☐ 0.3 ☐ 0.4 ☐ 0.5

19. Flexible working practices

181. How well are workers encouraged to become multiskilled? Do you have written learning programmes to facilitate multi-skilling? ☐ 0.1 ☐ 0.2 ☐ 0.3 ☐ 0.4 ☐ 0.5
182. How effective is your policy of recruitment to ensure high standards of education and aptitude, as a foundation for a flexible workforce? ☐ 0.1 ☐ 0.2 ☐ 0.3 ☐ 0.4 ☐ 0.5
183. How well have you removed the traditional barriers between engineering and production? ☐ 0.1 ☐ 0.2 ☐ 0.3 ☐ 0.4 ☐ 0.5
184. How successful is your workforce appraisal system in fostering movement of workers to higher grades? Is the process documented and freely available? ☐ 0.1 ☐ 0.2 ☐ 0.3 ☐ 0.4 ☐ 0.5

185. How effective are you in the use of temporary or contract labour for less skilled jobs? ☐ 0.1 ☐ 0.2 ☐ 0.3 ☐ 0.4 ☐ 0.5
186. How good are your performance-based reward schemes? Are they well documented and freely available? ☐ 0.1 ☐ 0.2 ☐ 0.3 ☐ 0.4 ☐ 0.5
187. How good is your team structuring with empowerment and designated team leaders? ☐ 0.1 ☐ 0.2 ☐ 0.3 ☐ 0.4 ☐ 0.5
188. How effectively are your team members trained in basic industrial engineering? (No training = 0.1, industrial engineer working with teams = 0.5.) ☐ 0.1 ☐ 0.2 ☐ 0.3 ☐ 0.4 ☐ 0.5
189. How good is your record for industrial relations in terms of hours lost due to industrial disputes? ☐ 0.1 ☐ 0.2 ☐ 0.3 ☐ 0.4 ☐ 0.5
190. How effective is your partnership with unions or representative labour bodies such as works councils? ☐ 0.1 ☐ 0.2 ☐ 0.3 ☐ 0.4 ☐ 0.5

20. Continuous learning

191. How effective is your on-the-job and off-the-job learning strategy? Is your policy documented and freely available? ☐ 0.1 ☐ 0.2 ☐ 0.3 ☐ 0.4 ☐ 0.5
192. How good are your learning programmes for implementing new technology? ☐ 0.1 ☐ 0.2 ☐ 0.3 ☐ 0.4 ☐ 0.5
193. How good are your written learning programmes for multiskilling and autonomous maintenance? ☐ 0.1 ☐ 0.2 ☐ 0.3 ☐ 0.4 ☐ 0.5
194. How good are your formal in-house education and learning programmes for managers and supervisors? ☐ 0.1 ☐ 0.2 ☐ 0.3 ☐ 0.4 ☐ 0.5
195. How well do you organize and implement learning programmes for a company-wide change process (e.g. TQM, TPM)? ☐ 0.1 ☐ 0.2 ☐ 0.3 ☐ 0.4 ☐ 0.5
196. How good are your learning resources in terms of full-time training staff, facilities and learning materials? ☐ 0.1 ☐ 0.2 ☐ 0.3 ☐ 0.4 ☐ 0.5
197. How well are employees encouraged and supported to further their general education at tertiary education institutions? Are general education opportunities limited to certain levels of employees? ☐ 0.1 ☐ 0.2 ☐ 0.3 ☐ 0.4 ☐ 0.5

198. How good is your learning performance in terms of training hours for each level of employee?

☐ 0.1 ☐ 0.2 ☐ 0.3 ☐ 0.4 ☐ 0.5

199. How good is your commitment to learning in terms of total training expenditure as a percentage of sales?

☐ 0.1 ☐ 0.2 ☐ 0.3 ☐ 0.4 ☐ 0.5

200. How successful are you as a learning organization in terms of awards and accredited programmes (e.g. 'Investor in People')?

☐ 0.1 ☐ 0.2 ☐ 0.3 ☐ 0.4 ☐ 0.5

10

Data recording and analysis

I have spent a fortune travelling to distant shores and looked at lofty mountains and boundless oceans, and yet I haven't found time to take a few steps from my house to look at a single dew drop on a single blade of grass.

Rabindranath Tagore

Some managers could be excused for being puzzled by the abundance of 'fashionable' concepts and the great variation of processes for benchmarking. Others might feel frustrated by the lack of standardization in benchmarking approaches.

We believe that there are substantial benefits to be gained from designing your own benchmarking process rather than following a consulting firm's standards. We are not suggesting that you should 'reinvent' the wheel, rather we explain how – by using the results from the 200 questions – you can tailor make your own benchmarking process. After all, every business is to some extent unique and each organization will have its own culture and a way of doing things. This culture should not be unduly upset. Any new process should consider a cultural fit.

In this chapter we discuss how to do your own benchmarking. We begin with suggesting how to go about collecting relevant data, and how to rate your organization against each question. This will enable you to analyse results so as to find the preliminary position of your business.

As Davey Crockett, of wild west fame, is reputed to have said, 'Be sure you are right then go ahead.' It is for you to choose what is right and then to drive ahead. The important success factors of conducting one's own assessment include:

- Employees know more about a company's business than any outside expert possibly can.

- A customized own benchmarking process will promote either a detailed or a less structured approach to fit the existing culture of an organization.
- Own benchmarking is likely to generate an ownership for a solution and dispel any distrust towards glib 'off-the-shelf' approaches.

Outside consultants can bring in experience, expertise and can be objective. They can often act as a catalyst to make things happen. They may also provide support to under-resourced organizations. None the less our recommendation is that you will gain the most benefits by first carrying out your own self-assessment. The experience of Johnson and Johnson in the early 1900s, as reported by Biesadra (1992), was that by relying heavily on consultants the company initially achieved only 5 per cent of what was finally achieved through internal benchmarking activities. Costs and cycle times of achieving results were greatly reduced when the strategy changed from using external consultants to own employees. The quality of results were also greatly improved.

Our recommended steps and methods for carrying out self assessment are reviewed in Chapter 13.

Data preparation

Following your decision to embark upon a benchmarking project the next stage is to organize a multi-functional team. Team members should be senior and experienced people and all functions of the business should be represented. The first step for this team is to understand the company's organization and business activities. It is likely that the background and experience of team members will be varied and many of them may not have a clear understanding of the total business. People from engineering and technical backgrounds may understand detailed procedures of technical operations, while people from sales, marketing and service backgrounds may be more conversant with broader issues related to customers.

The synergy of the team should be greatly enhanced by comparing each member's understanding of the company with regard to the following basic questions:

- What is the core business of the company?
- Who are the main customers?
- Who are the top three competitors?
- Which sectors of the business are the most and least profitable?
- What is the size of the business in terms of sales, assets and people?

DATA RECORDING AND ANALYSIS   191

Although agreement on the answers to the above questions will promote a synergy for the team, the 200 questions cannot be considered until substantial amounts of data are collected and in some cases collated. The detailed preparation of data, and the understanding of the business process, can largely be derived from the following basic information as it exists:

1. Annual operating plans for the last five years showing actual sales trends and profitability by product.
2. Long-term strategic plans, or study reports, giving market forecast by products and sectors.
3. Internal company reports and databases on competitors' market share and innovation programmes.
4. Current information on customers and distribution channels.
5. Expenditure programme (budget) for the next five years regarding innovation, marketing and capital projects.
6. Written policies and programmes on product safety, industrial safety and environmental protection.
7. Non-financial performance measures regarding:
   - customer service
   - plant efficiency
   - plant utilization
   - maintenance effectiveness
8. Quality programmes and ISO 9000 certification if appropriate.
9. Information technology policy, and existing hardware and software application programs.
10. Organizational structures, management development and learning programmes.
11. Annual financial reports.
12. Marketing research reports and databases.

For the team to gain an in-depth understanding of the total process further analysis will be needed. One useful way to describe and analyse a business is by process mapping or process flow diagrams. This approach is illustrated in Figure 10.1. However, the description and analysis of the business may not need to be as complicated as this illustration. Once the business process is mapped, the team possibly might identify some areas of bottlenecks before carrying out a detailed self-assessment.

The next stage for the team is to study and understand the six pillars and the twenty foundation stones as described in Chapters 3 to 8. It is vital for the team to review all aspects of the 20 foundation stones so as to identify and understand which factors are critical to achieving excellence for your business. The well-known value-chain model of Michael Porter (1990, see also Chapter 4) is a useful way of aligning your business

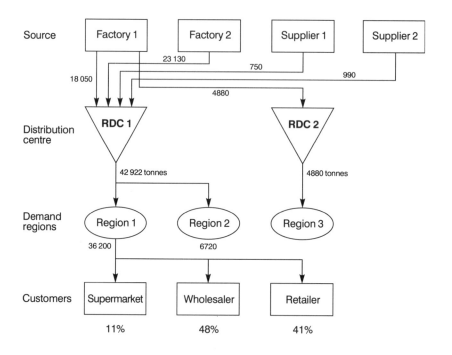

Figure 10.1 *Flow diagram of sourcing and distribution of a product*

with the six pillars in order to comprehend their characteristics. Figure 10.2 shows a model of our six pillars based upon Porter's value chain. We have three primary pillars of performance activities (marketing and innovation, manufacturing facilities and supply-chain management). They are supported by the three secondary pillars of practices (environment and safety, procedures and people). The combination of performance and practices gives excellence as depicted in our model.

As in Porter's model, one can focus the business needs by analysing the relative value of each pillar and the reason for its existence with regard to the business. In our model the pillars of primary activities are profit driven and the aim is the achievement of higher performance standards. The pillars of secondary activities are service driven and the aim is for achieving best practices.

Although not essential, it is helpful to obtain published reports on out-

DATA RECORDING AND ANALYSIS 193

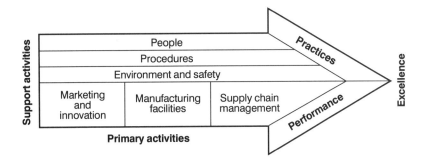

Figure 10.2 *Value chain. Source: Porter, 1990*

standing or 'world class' performance achieved by other companies in your area of interest. We do not recommend that you should hunt for 'world class' companies as suitable benchmarking partners. A wealth of information is available in published reports and this should be sufficient to provide a better understanding of what is achievable. Several 'best practice' databases are now available in the USA and Europe which will enable you to assess against the reputed best.

Guide to assessment

A potential danger of performing your self-assessment by means of our 200 questions lies in its deceptive simplicity. Your own benchmarking is not conceptually complicated but it is extremely rich in detail. We have not applied any psychological expertise in the formulation of our questions, but it is evident that without adequate guidelines your teams could be at a disadvantage. It is often tempting to 'guess' a rating if the basis for the rating of a specific question is not properly qualified. In Chapter 9 we have used a rating method for our questions.

Bogan and English (1994) give four categories of questions for this type of survey, as follows:

1. Multiple choice
Example: What type of maintenance policy do your engineers follow?
a) Breakdown
b) Time-based preventive

c) Predictive
d) Inspection based?

2. Forced choice
Example: Are you satisfied with forecasting methods and techniques?
Yes ☐
No ☐

3. Open-ended
Example: What suggestions do you have to improve your customer service?

4. Scales
Example: Rate on a scale of 1 to 5 with 1 being 'poor' and 5 being 'excellent'.
How good is the success ratio of new products?

The difficulty with multiple choice questions is that they provide a set choice of answers and sometimes the choice given does not adequately fit the situation. With forced choice (yes/no questions), no shading at all is provided. Seldom will answers be black or white.

With open-ended questions, however, too much leeway is given, a large amount of subjectivity is allowed and frequently answers stray from the question. In addition sometimes respondents find difficulty in replying to open-ended questions if they have to prepare a written answer.

We decided that the scaled method of questions would provide the best method for our purposes. The scale has a sufficient span to permit a reasonably accurate rating by an experienced team. Black and white type answers are avoided and subjectivity is confined within the limits of the five-point scale.

Some questions relate to areas where data is measurable (e.g. question number 135 regarding plant efficiency). In this case it is relatively easy to apply a rating scale, provided a range of performance level can be defined for each scale. But it has to be clearly understood that it is essential to have a good understanding of what best practice is, that is what exactly is the standard we are trying to measure against. A 'complacent' rating of 0.4 and 0.5 for every one of the 200 questions will render the exercise meaningless. The other danger is that once a rating is given it takes on an aura of accuracy which may be entirely misleading unless the rating has been given with full knowledge of all factors by an experienced team.

As many questions do not lend themselves to a quantifiable answer, the answers by necessity will rely on interpretation by team members. We recognize that interpretation may give scope for some inconsistency by teams. It is our intention to develop a piece of 'knowledge-based' software incorporating guidelines for all 200 questions. In the meantime, as a guideline, the following indicative samples are provided.

Question 1

How well do your managers in marketing and sales know the relative importance of main products (by volume, profit and trends)?

Poor (0.1)

No measurement of sales volume, value or profit by product. Total sales are usually reported by region or salesperson.

Fair (0.2)

Some measurement by product of sales volume, value and profit. No trend analysis. Marketing and sales managers do not receive regular reports by products.

Good (0.3)

Regular reporting of sales volume, value and profit by products. Trend analysis is carried out. Generally marketing and sales managers are reasonably competent in the interpretation of figures.

Very good (0.4)

Good measurement and reporting of sales volume value and profits by products. Regular trend analysis. A computerized information system facilitates ABC analysis by various parameters. Marketing and sales managers are trained and able to take actions from regular trends and performance reports for each product.

Excellent (0.5)

Very reliable and prompt measurement and reporting of sales volume, value and profits by product. Trends are plotted on computer systems which are accessible by sales and marketing managers. Managers are capable of analysing data (e.g. ABC analysis) and take action on an interactive real time basis.

It would have been ideal to list detailed guidelines for all of the 200 questions, but that would create an imbalance in the book with relatively low added value. This apparent deficiency can be redressed by adequate training and exchanging assessment of 'non-quantifiable' questions between team members. We also propose to develop software containing guidelines.

Assessment and scoring

Verification of data

Following data preparation and preliminary data collection the team should verify key information through a series of carefully selected interviews. One simple rule the team must follow is open communication. People being interviewed, or providing information, have to be reassured that there is no ulterior motive. It has to be made clear that the objective is to carry out a 'health check' of the total business and not of any one person or department.

Agreement of scores

After following the hints given above ('Guide to assessment') it is useful for each member of the team to carry out a 'dry run' in order to ensure that all questions can be adequately answered. The team can then have an open discussion to exchange views on doubtful areas. The next stage is relatively straightforward with each member of the team, without collaborating with other team members, ticking off their assessment of appropriate scores against each question. Finally the team will meet together to review all answers and to agree a set of scores.

We emphasize that regardless of the type or priorities of a manufacturing business, the first assessment must include answers to all 200 questions. No question should be given a 'nil' score. In Chapter 12 we show how the scores from a first assessment can be recalculated according to the company mission and business objectives.

Data analysis

The object of data analysis is to bring order out of the scores for the 200 questions. Data analysis can be done in a number of ways. We recommend four simple but effective means of data analysis:

- Rating profile
- Manufacturing correctness factor (MCF)
- 'Spider' diagram
- Histogram

Rating profile

As illustrated in Figure 10.3, a rating profile is drawn simply by joining scores of all 200 questions. It provides two useful pointers:

- To ensure that all questions have been assessed with scores in a scale 0.1 to 0.5.
- To highlight a rough guide to the 'health' related to each question. Other diagrams show performance measures by a foundation stone and not by individual questions.

1. Understanding the marketplace

1. How well do your managers in marketing and sales know the importance of main products?
2. How good (precise) are your analyses of trade needs and consumer habits?
3. How often do you conduct market research of trade needs and consumer habits?
4. How often do you evaluate product performance in the market?
5. How well do your sales and marketing team know the relative importance of functions that affect customer satisfaction?
6. How systematic and sophisticated are your advertising and promotion activities?
7. How close is the link between your sales, marketing, planning and manufacturing functions?
8. How good is your manufacturing managers' first hand knowledge of the market place?
9. How well do you know international tariff tax and trade regulations?
10. How aware are you of opportunities and constraints for emerging markets?

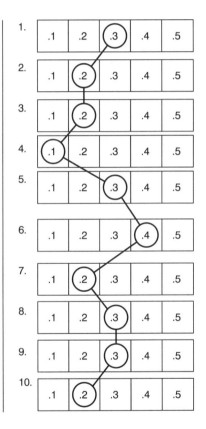

Figure 10.3 *Example of a rating profile*

Manufacturing correctness factor (MCF)

MCF is the aggregate total of scores expressed as a number, the maximum possible value being 100. The scores for each foundation stone are added and then summarized as shown in Table 10.1 to calculate the MCF.

Table 10.1

	Foundation stone	Score
1	Understanding the marketplace	5
2	Understanding the competition	4
3	Product and process innovation	4
4	Manufacturing resources planning and working with suppliers	4
5	Distribution management and working with customers	4
6	Supply chain performance	5
7	Product safety	2
8	Industrial safety	3
9	Environmental protection	3
10	Sourcing strategy	4
11	Appropriate manufacturing technology	5
12	Flexible manufacturing systems	3
13	Reliable manufacturing	4
14	Manufacturing performance	4
15	Quality management	3
16	Financial management	3
17	Information technology	2
18	Management skills and organization	3
19	Flexible working practices	2
20	Continuous learning	3
	Manufacturing correctness factor	70

'Spider' diagram

MCF is a new concept and it provides a useful indicator of the overall performance of the business. But it does not highlight the strengths and weaknesses of individual foundation stones. The purpose of a spider diagram is to do just that. It has been used before in some modified forms and under various names such as 'arachnid' diagram, radar chart, measured matrix chart or M2. As illustrated in Figures 10.4, the chart consolidates various performance scores by arraying different foundation stones along the radius or spokes of a circular graph.

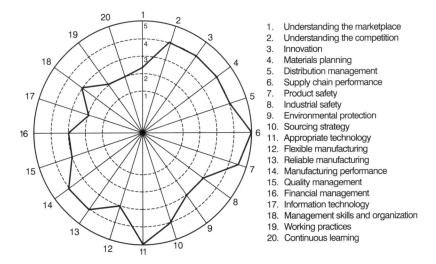

Figure 10.4 *Manufacturing correctness profile ('spider diagram')*

Greatest opportunities for improvement are highlighted by the size of gaps between the centre point and the radius. Different target goals, appropriate for each foundation stone, can be set on the diagram to identify the target gaps. We shall discuss this with examples in Chapter 13. The 'spider' diagram offers a number of positive features including:

- It complements the manufacturing correctness factor by highlighting how MCF is made up.
- It illustrates the total business performance levels with a single graph.
- It pinpoints the performance gaps for each foundation stone.
- It focuses the management attention on total manufacturing solutions rather than on isolated performance figures produced for a specific area or foundation stone.

Histogram

A histogram has the same objective as a spider diagram, i.e. to illustrate the performance levels of all Foundation Stones in a single graph. The difference is that a histogram shows linear rather than radial results.

Some users have found a histogram, as illustrated in Figure 10.5, visually easy to interpret and to measure gaps with respect to targets. The target or best practice figures can be presented alongside the actual performance for each foundation stone as shown in Figure 10.6.

200 TOTAL MANUFACTURING SOLUTIONS

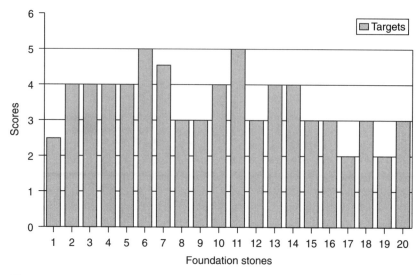

Figure 10.5 *A histogram of scores for foundation stones*

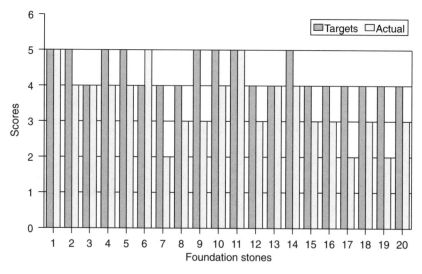

Figure 10.6 *A histogram of scores for targets versus actual*

Computer-aided analysis

We all know that effective computer systems are good enabling tools for successful projects. Software should be able to assist teams particularly at the data preparation stage of the project. Some examples of available software applications are:

- Spreadsheet programs (e.g. Excel, Lotus 1-2-3) to help teams store data, calculate performance figures, and carry out trend analysis.
- Process mapping programs to enable teams to prepare flow charts of various business processes.
- Graphics programs (e.g. Freelance, Power Point) to assist teams display data and results.
- Word-processing programs to prepare questionnaires, interim reports, etc.
- CD-ROM libraries, Lotus Notes and databases to carry out literature searches and for best practice examples.
- E-mail systems to communicate between team members and key people.

It is not essential to hunt for all the software options. The use of existing technology with which the team members are familiar should be adequate for their preparation work. However it would be extremely useful if the team had access to specially designed software to:

- Provide guidelines for assessing the answers of all 200 questions.
- Compute the total score of each foundation stone and thus determine the manufacturing correctness factor.
- Produce spider diagrams and histograms.
- Include flexibility such as user-defined questions.

We have developed such a software on Microsoft Windows which meets the above user requirements and named it ASK (Advanced Self Analysis Kit). It is designed to incorporate guidelines to 200 questions and to enable users to 'self appraise' themselves against world class standards.

Performance versus practice analysis

The inclusion of performance versus practice analysis was prompted by the success of a benchmark study carried out jointly by IBM and the London Business School and reported by Hanson et al. (1994).

Continuing the value chain approach, as shown in Figure 10.2, we can classify the six pillars and their associated foundation stones into two groups:

(a) **Performance pillars (primary activities)**
  1. Marketing and innovation (foundation stones 1-3)
  2. Supply-chain management (foundation stones 4-6)

3. Manufacturing facilities (foundation stones 10-14)

(b) Practice pillars (support activities)
4. Environment and safety (foundation stones 7-9)
5. Procedure (foundation stones 15-17)
6. People (foundation stones 18-20)

There are 11 foundation stones for the performance pillars accounting for a maximum score of 55. Similarly the nine remaining foundation stones constituting practice pillars contain a maximum score of 45.

In order to achieve and sustain a leading competitive advantage, a manufacturing company must show 'very good' or 'excellent' results in both performance and practice indices. There are companies who score well primarily in 'performance' based primary activities. Other companies achieve excellent results in 'practice' parameters, perhaps through a quality management approach, but fail to do well in 'performance' parameters. A number of organizations have the potential to quickly improve their status but a large group is lagging behind. We have therefore divided manufacturing companies into five groups based on their practice and performance scores as illustrated in 'practice performance map' (see Figure 10.7).

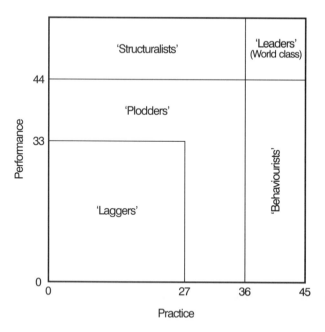

Figure 10.7 *Practice performance map*

Leaders ('world class')

These are the companies achieving high scores both in practice (36 to 45) and performance (44 to 55). The manufacturing correctness factor (MCF) for these companies will be over 70. These companies are capable of competing with the best of the world's manufacturers.

Behaviourists

These companies have a high score in practice (36 to 45), but relatively low scores in Performance (less than 44). Their total MCF will range between 47 and 84. These companies are promising because they have the supply infrastructure and the change culture. With a structured improvement programme towards higher performance levels they can reach the 'leaders' category.

Structuralists

These are organizations with a high score in performance (44 to 55) but a lower score in practice (less than 36). Their MCF values will range from 53 to as high as 90. These companies appear to be ahead in the game, but in the longer term they are not likely to sustain their performance advantages. In order to move towards the 'leaders' category they will need to invest in time and resources in best practices and training.

Plodders

The fourth group of companies possess medium scores for both practice (27 to 35) and performance (33 to 43). Their MCF values could be between 42 and 78. These companies may have potential but they lack both performance and practices to compete in the world stage. They will require a long-term improvement programme with significant changes to their company policy, operation and practices.

Laggers

The lowest scoring group in both practice (less than 27) and performance (less than 33) constitute the category of laggers. We need say no more.

Conclusion

This chapter has discussed how initial data for analysis and rating can be drawn from information that already exists in the form of reports, plans and databases. A method of data analysis and our method of rating is detailed. Knowing where the data is, and how to rate it is one aspect. But perhaps the more important issue is the formation of a team to carry out the rating. The team will need to be drawn from senior people in each function.

The team will have to familiarize themselves with the complete structure and processes of the organization. Just by doing this the team members will become more valuable members of the organization! Team members will also have to read and understand Chapters 3 to 8 so as to understand what constitutes best practice, what is possible, and what to look for. In their quest to understand the organization, team members will doubtless have to talk and question people in each department as to what actually happens. Likewise when the rating process is being carried out team members will again have to refer back to the people who are actually doing the job and who are making things happen, to find out what and how things are actually done. It would be surprising if all this fact gathering and question asking didn't cause speculation by the people of the organization. It will therefore be absolutely essential for team members to be completely open in what they are trying to achieve.

The aim is to improve the health of the organization and not to blame or to censure any one person or group of people.

The gathering of information, the familiarization of team members with the organization and processes of the organization, and the gaining of the understanding of what is possible and what to look for might seem to be a painstaking project. We would have to agree it won't be easy, but the results will be far reaching. In the long run we know our approach is in fact a lot simpler and easier to implement than you might first imagine.

References

Biesadra, A. (1992) Strategic Benchmarking. *Financial Week*. September 29.
Bogan, C.E. and English, M.J. (1994) *Benchmarking for Best Practices*. McGraw-Hill.
Hanson, P., Voss, C., Blackham, K. and Oak, B.J. (1994) *Made in Europe – A Four Nations Best Practice Study*. London Business School.
Karlof, B. and Ostblom, S. (1994) *Bench Marking. A Sign Post to Excellency and Productivity*. (Trans J. Gilderson) John Wiley and Sons.
Porter, M., (1990) *The Competitive Advantage of Nations*. Macmillan

11
Mission statements

> *The Moving Finger writes; and, having writ,*
> *Moves on; nor all thy Piety nor Wit,*
> *Shall lure it back to cancel half a Line,*
> *Nor all thy Tears wash out a Word of it.*
> The Rubaiyat of Omar Khayyam
> (trans. E. Fitzgerald)

In previous chapters we have stressed the need to strive for quality and the elimination of non-value-added activities. In Chapter 8 we considered the 'people' pillar of manufacturing correctness and the importance of cultural fit. We concluded that no business can hope to become a world class organization unless all the people in the organization know what is required and are sufficiently enthused to make it happen.

In this chapter we discuss how to get the message across and how to harness the enthusiasm of the people of an organization by using a mission statement. We show that if properly constructed and worded, a mission statement will communicate to those within the organization where the organization is heading (the vision) and how it is going to get there (the strategy). On its own, no matter how carefully worded, a mission statement will not achieve anything unless it is in tune with what the people of the organization want to happen. The mission statement has to reflect the values of the people. In many cases the values of the people, or the culture of the organization, will need to be changed if the quantum leap to world class manufacturing is to happen. A carefully crafted mission statement is an essential element in changing the culture of the organization. For, as earlier stated, unless the people want it to happen it won't happen.

Quality is everybody's business. Churchill once said that war is too important to be left to the generals. So too with quality; quality is too important to be left to the managers. Everyone in the organization has to be involved, and everyone, in the words of Tom Peters (1986), must have a passion for excellence. But, as discussed in earlier chapters, with the value chain approach it is not only the managers and the members of the organization who should be passionately involved in quality and the elimination of non-value activities, but *suppliers* are also expected to be imbued with the same enthusiasm. Likewise, if *customers* can be involved in advising and specifying what changes or improvements they would like, they too become part of the value chain and consequently are expected to be an integral part of the quality culture. Quality then is the concern of all those involved in the total value chain, beginning with the suppliers and flowing right through the process to include the customers. There is one school of thought that would suggest that the value chain actually begins with the customer on the grounds that the customer is the first input into the process. This would certainly be true of Toyota and the 72 hour car concept referred to in Chapter 2.

Thus in framing a mission statement not only will the aim be to give the people of the organization the vision and a reference point for strategy, it should also be a statement to suppliers and customers of the values or culture of the organization.

We conclude this chapter with the constituents of an effective mission statement.

Culture

In a centralist, or bureaucratic, organization the culture is generally that some people, the minority, do the thinking and give the orders and other people, the majority, receive orders and under close supervision wield the screwdrivers and do what they are told. In these types of organizations a standard of quality is set from above, and up to a point conformance is achieved by inspection, testing and checking. The quality achieved in this type of culture is generally at a low level just sufficient to pass basic inspections (near enough is good enough). The lower level workers are not expected or encouraged to make suggestions or to question and it is expected by the inspectors and supervisors that there will be a percentage of sub-standard work. There is little encouragement for workers to make suggestions to improve the quality level as bare conformance to the lower level of the standard is sufficient. The different approaches to quality are discussed in Chapter 7.

But an organization that embraces total quality management is on a different plane. Such an organization is concerned with more than just standards and conformance; such an organization has an overriding culture of excellence and quality. Excellence and quality mean more than being customer focused. Excellence and quality also mean that everyone in the organization, from the chief to the cleaner, is determined to eliminate any cost that does not add value to the process or service. With a total quality management culture every member of the organization seriously believes that not one day should go by without the organization in some way improving the quality of its goods and service.

Vision

A culture where everyone in the organization has a passion for excellence begins with a vision. The vision must begin with the chief executive.

But it is no use if a leader has a vision which is not communicated to the rest of the organization, or if the rest of the organization does not want or is not interested in making it happen. For example as Albrecht (1988) puts it, what do you do if the elephant does not want to dance?

Top management has to sell the vision to the rank and file. Once the rank and file are won over then they will become the driving force for quality. Once the culture of quality has been firmly entrenched within the organization it will permeate outwards to embrace suppliers and customers. Once this happens management will no longer be attempting to dictate the level of quality nor directing how the level might be achieved. Customers, suppliers and in-house lower-level workers will all be involved in making daily incremental improvements and giving suggestions to management for larger, far-reaching improvements. Thus the drive will now be bottom up rather than dictated from above. Everybody in the organization will be involved in sharing the same vision of getting it right first time and always with a customer focus. This may seem a very simple concept and indeed it is.

However, the introduction of total quality management requires an understanding by management of what is entailed, i.e. full commitment of management, careful planning, intensive training, and a change (often drastic) in culture at all levels of the organization. Total quality management cannot be achieved without the vision and everyone must buy into the vision. The elephant has to be persuaded that it wants to dance.

Vision and cultural fit

The word vision suggests almost a mystical occurrence (Joan of Arc), or an ideal (such as expressed by Martin Luther King, 'I have a dream'). The same connotation is found when looking at a vision in the organizational context. A leader with a vision is a leader with a passion for an ideal. Nanus (1992) says that there is no more powerful an engine for driving an organization than an attractive worthwhile vision of the future widely held. His definition is that 'a vision is a realistic, credible, attractive future for your organization'. He adds that 'the right vision is so energizing that it in effect jump starts the future by calling forth the skill, talents, and resources to make it happen'.

El-Namaki (1992) also stresses future reality. In this he follows Polak (1961) who says that vision is where tomorrow begins, for it expresses what you and others who share the same vision will be working hard to create. Polak uses great visionaries such as Moses, Plato and Karl Marx to illustrate his point.

> Themselves under the influence of what they had envisioned, they transformed the non-existent into the existent, and shattered the reality of their own time with their imaginary images of the future. Thus the future always operates in the present, shaping itself in advance through these image makers and their images.

Polak defines vision as a 'concept for a new and desirable future reality that can be communicated throughout the organization'. But unless the vision can happen, it will be nothing more than a dream. As Langeler (1992) observes 'grand, abstract visions may be too inspirational. The company may wind up making more poetry than product.'

As Stacey (1993) says, 'The ultimate test of a vision is if it happens.'

Culture change

To make a vision happen within an organization there has to be a cultural fit. Corporate culture is the amalgam of existing beliefs, norms, and values of the individuals making up the organization ('the way we do things around here'). The leader may be the one who articulates the vision and gives it legitimacy, but unless it mirrors the goals and aspirations of the members of the organization at all levels then the vision will not happen. 'Once expressed the vision may be killed if it goes against basic beliefs' *El-Namaki* (1992).

Culture and values are deep seated and may not always be obvious to the members themselves. As well as the seemingly normal aversion to change by individuals, often there is a vested interest for members of an organization to resist change. Middle management is often more likely to resist change than other members. In 1513 Machiavelli wrote, 'It must be considered that there is nothing more difficult to carry out, nor more doubtful to success, nor more dangerous to handle, than to initiate a new order of things.' Human nature hasn't changed much since the sixteenth century!

Remember that the organization does not have a brain. Organizations are made up of many individuals, each with their own set of values. The culture of an organization is how people react or do things when confronted with the need to make a decision. If the organization has a strong culture then each individual will instinctively know how things are done and what is expected. Conversely, if the culture is weak, then the individual may not know how to react and thus can react in a manner contrary to what management would hope.

How then can a cultural fit for total quality management be engineered?

Leadership and mission statements

First of all there has to be leadership from the top. Leading by example might be a cliché but, unless the chief executive can clearly communicate a clear policy, how will the rest of the organization know what is expected? We strongly recommend the mission statement as a starting point for the communication and the gaining of acceptance for a change of direction.

Mission statements

Today there are few companies that don't have a mission statement.

If the word 'vision' has almost mystical, certainly inspirational connotations, then the more pragmatic word 'mission' suggests getting on with the job. For example if a church leader had a vision of converting the world to the 'true' faith, and if the leader was able to imbue others with the same enthusiasm, then visionaries were sent out with the mission of making the vision happen. These visionaries (missionaries) would have a clear cut mission, that of taking active steps to transform the vision into reality.

Or, as expressed by Dulewicz, MacMillan and Herbert (1995):

> A vision depicts the aspirations of the company, a desired and attainable picture of how the company will appear in a few years' time, which can capture the imagination and motivate employees and others. The mission is to achieve the vision, expressing the commitment and will to do so. On the way, decisions will have to be made according to the values of the company, as indicated in the decision-making behavior of the board – according to what the board believes is good or bad, right or wrong from the company's point of view.

Research by Coulson-Thomas (1992) revealed that new chief executives usually seem to feel it is mandatory to issue a new mission statement. The ostensible reason is to communicate a new mission, or a change of direction. In this sense the mission is given as a statement of where we are going (the vision) and how (the strategy) we are going to get there. But as Klemm, Sanderson and Luffman (1991) found, there is 'a weight of evidence' to show that often the real reason for the new statement is for the new chief executive to establish authority, and to show that the organization has a new leader (the 'I'm the boss and things are going to change around here' syndrome).

Coulson-Thomas (1992) also found, from a series of four surveys of British companies, that in reality mission statements do not reflect the true situation.

> In most companies there is a feeling that missions are just words on paper. Directors and senior managers are not always thought to be committed to their application.

An example is given of a managerial interviewee who pulled a mission statement out of his wallet saying 'Here it is. They put it on a card. I couldn't tell you what it says. It's one of those things that doesn't stick, but we've all got one'. Another director is quoted as saying, 'I bought into quality, and the mission/vision as did my team. Everyone talks quality but I'm measured by the same old ratios and it (the mission) is great but the numbers are real.'

Likewise, from a study of over 200 mission statements (Wright, 1995) the conclusion reached was that most mission statements are mere rhetoric, full of sound and fury and signifying nothing. Often used phrases include 'to be the best', 'to be world leaders', 'to offer first class, or world class quality service', and so on, and if the names of the organizations were taken away most statements would be interchangeable without anyone being the wiser! From an interview of 50 middle managers drawn from five different companies it was found that few could remember what the mission statement for their company said, and in the few cases where managers could roughly recall what the statements said they couldn't explain what was meant.

If the mission statement is mere words and if managers can't remember what the statement says then, although it might give the writer a nice warm feeling, it is unlikely that others will be motivated to change their attitudes or the ways in which they do things.

A believable mission statement has to be honest. Workers are not fools, hidden agendas soon become transparent. When we hear a chief executive say people are our greatest resource, cynically we are not surprised when soon afterwards restructuring and redundancies are announced. The same is true of mission statements; a mission statement that consists of platitudes will not generate enthusiasm. Grandiose or naive missions are self-destructive. Mission statements must be honest.

There is another aspect to a badly written mission. As far back as 1960,

Levitt was saying in his article 'Marketing Myopia' that most companies had the wrong business definition. He reasoned that a railroad company should see its business as moving people and goods rather than just railroading. His contention was that once the business, rather than the technology was defined, then management could focus on customer needs rather than on production technology (Levitt, 1960).

Getting effective decision making down to the lower levels of the organization requires the creation of the right structure so that people can get involved, can become committed and believe that they are able to make things happen. Real leaders communicate face-to-face not by memos. The flatter the structure, the closer the leader will be to the true workers, that is, the people who actually are involved in adding value in the process, rather than those who are administering and regulating. A real leader creates leaders whereas a manager tries to retain control and in doing so creates, or perpetuates, a culture of compliance and conformance. As discussed in Chapter 8 the transformation from manager to leader involves a major paradigm shift.

To effect a change, there has to be leadership and a clear statement by the chief executive of exactly what is expected.

Change requires careful planning, harmonious collaboration, and a willingness to listen and to accept criticism and suggestions. The first step for a chief executive will be to win the board over to the understanding that quality will give the competitive edge. If an organization cannot see that quality is important then they will make life very easy for their competitors. In a global marketplace there is nowhere to hide, no organization can hope to ignore the competition. Once the board accepts that we are all operating in a highly competitive global market and that quality will give the competitive edge, then the senior managers have to be convinced. Until there is whole-hearted agreement and a determination at board and senior management level, then it will not be possible to sell, or internally market, the quality message and the necessary changes in attitudes to the lower echelons.

We believe that if the organization as a whole is going to take ownership of a mission statement, ideally each member of the organization should be given an opportunity to be involved in the formulation of the statement. As the aim of the mission statement is to make the vision happen, and as the vision begins with the chief executive, it follows that the first draft of the mission statement must come from the chief executive's office. The draft then should be circulated and tested on the board, company employees, suppliers and customers, in short all the stakeholders. This will permit all involved or connected with the organization to provide input and it will certainly test if the mission statement is in tune with what they want to happen.

Schein (1988) points out that any change process involves not only learning and believing in something new, but unlearning something that is already present. Thus, no change will take place unless there is a motivation to change and the need for change is fully understood. All changes have to be negotiated, that is agreed to, before a stable change can take place.

Constituents of an effective mission statement

We have concluded that if a mission statement is to be effective in helping a drive to quality and customer focus then the following constituents are required:

1. The chief executive has to have a *vision* and be passionate to make the vision happen.
2. The mission statement must be specific. Generalizations must be avoided.
3. The mission must fit the culture of the organization.
4. The mission has to be honest (staff will soon realize that there is a secret agenda well before the redundancies are announced).
5. Input from all stakeholders (the board, suppliers, customers and the people of the organization at all levels), in formulating a statement will help to foster a sense of ownership.
6. Once formulated the mission must be communicated and must be conspicuously available.

Cementing the change

A change to a quality culture will not happen overnight, but unless the above steps are taken it will not happen at all. Just as important as getting commitment to change will be the actions taken to cement the changes. As underlined in Chapter 8, a change to quality is not only for the factory workers. The biggest changes in attitude often will have to be made at the top of the organization. Likewise often middle management will feel the most threatened.

Once the quality revolution has taken hold then executive privileges will become less important. A leader leads by example. A leader does not need a separate office; a true leader will want interaction and will want to be where the action is. The action is not to pore over figures and budgets and draw up new mission statements to establish authority: the action is in the front line.

It appears to be a world-wide phenomenon that some senior managers will opt for early retirement or will move on rather than accept the changes necessary to bring about a total quality management culture. Change and the 'giving up' of power will be too difficult for them to handle. As the change rolls on down through the organization, additionally it will be found that some middle managers and quite a few supervisors will also opt to leave. The problem is that we have become used to the hierarchical model which requires some people to give orders and other people to take orders. This has caused a mind set where it is hard for people to give up power and to trust the lower echelons to get things right. Some, too, will find it difficult to give up the trappings of power. The company car, the parking space and the private office are all treasured and are the outward evidence of power and success not only within the organization but to friends and family.

But it is also just as difficult for those who have been used to receiving orders to accept responsibility. Some people prefer to be told what to do and are not comfortable with making decisions and accepting responsibility. Empowerment is a two way thing – managers have to be prepared to let go and trust, and workers must be prepared to accept responsibilities.

With a quality culture, there is no room for people or for expenses that do not add value to the process. It is best to let people go who do not want or who cannot change. This will be one of the harder decisions that will have to be taken.

Summary

A change in culture to total quality management involving customers and suppliers will not happen without careful planning and the enthusiastic commitment of everyone in the organization. This chapter has shown the importance of involvement of all the stakeholders (the board, the people of the organization at all levels, plus suppliers and customers) in crafting a mission statement that all can believe in. It is concluded that an honest statement in tune with the stakeholders' values will gain the enthusiastic commitment of all and will be the catalyst for the change to a quality culture.

War is too important to be left to the generals, quality is too important to be left to the managers.

References

Albrecht, K. (1988) *At America's Service.* Dow Jones-Irwin.
Coulson-Thomas, C. (1992) Strategic Vision or Strategic Con? *Long Range Planning,* 25, No. 1, 81-99.
Dulewicz, V. MacMillan, K. and Herbert, P. (1995) Appraising and Developing the Effectiveness of Boards and their Directors. *International Journal of General Management,* 20, No. 3, Spring, 1-19.
El-Namaki, M.S.S. (1992) Creating a Corporate Vision. *Long Range Planning,* 25,

No. 6, 25-29.

Klemm, M., Sanderson, S. and Luffman, G. (1991) Mission Statements: Selling Corporate Values to Employees. *Long Range Planning*, 24, No. 3, 73-78.

Langeler, G.H. (1992) The Vision Trap. *Harvard Business Review*, March/April, 46-49.

Levitt, T. (1960) Marketing Myopia. *Harvard Business Review*, July-Aug, 45-56.

Machiavelli, N. (1513) *The Prince*. (Trans., Luigi Ricci, revised by E.R.P. Vincent), New American Library of World Literature, 1952: New York.

Nanus, B. (1992) *Visionary Leadership*. Jossey-Bass.

Polak, F. (1961) *Image of the Future*. (Trans., E. Boulding), Volume 2, Amsterdam: A.W. Sijhoff.

Schein, E.H. (1988) *Organizational Psychology*. Prentice-Hall: Englewood Cliffs, N.J.

Schein, E.H. (1991) *Organizational Culture*: A Dynamic View. Josey Bass: San Francisco.

Stacey, R.D. (1993) *Strategic Management and Organizational Dynamics*. Pitman: London.

Wright, J.N. (1995) Creating a Quality Culture. *International Journal of General Management*, Winter.

12

Gap analysis

Nothing has the power to broaden the mind as the ability to investigate systematically and truly all that comes under thy observation in life.
                                                                Marcus Aurelius

In previous chapters we have dealt with the six pillars and the 20 foundation stones which are the underpinning manufacturing success factors. Two hundred questions based upon the specific criteria of each foundation stone have been provided to assess the whole 'structure' of your business. The mission statement (Chapter 11) has shown the importance of involvement of all the stakeholders in the business. In addition the mission statement has focused true business priorities. Now is the right stage to take the most important step of analysis, that of the gap analysis. The purpose of the gap analysis is to study the performance gaps as shown in Chapter 10 and to identify the underlying factors which explain the causes of the gap. With accurate measurements and systematic analysis it will become easy to pin-point areas for attention. The improvement of results for the whole organization is the ultimate goal, and gap analysis is a powerful approach to achieve these results.

The key steps of the analysis include establishing the targets for performance or best practice, normalizing the measured performance levels achieved, analysing the performance gaps and their causes by using simple graphical tools and developing an improvement strategy. If the study does not lead to a practical improvement plan, it would be a waste of effort and all the charts would merely be decorative wallpapers.

Setting targets

For total manufacturing solutions to have a real impact, in general a company should set its sights on the 'leaders' position. If the outcome of your self-analysis indicates your organization is either a 'behaviourist' or a 'structuralist' then going for gold in all weak areas may be justifiable. But the 'search for the holy grail' in each foundation stone may take the 'laggers' forever. Bogan and English (1994) advise, "The holy grail-like search will make you an 'industrial tourist' but not an effective benchmarker".

Therefore, after positioning your business from the scores of self analysis, the rational next step is to modify or confirm your targets in accordance with your mission statement and business objectives. You need to understand beyond the rhetoric in your mission statement as to what critical factors in your business must be altered to enhance your competitive edge. You may have a low score in a particular foundation stone (say, product safety) but if your business priority depends on this foundation stone then you should set a higher target.

We suggest that you should establish your targets at two levels. First, each foundation stone should have an average target score of 0.4 or 0.5. Only if a foundation stone is not relevant to your business, and this will be rare, then a target value lower than 0.4 may be considered. By default the target score is 0.5. The second level of target involves the quantitative measures for selected questions. For example, you may set a target of distribution cost as 2 per cent of sales for question number 59.

If your manufacturing correctness factor is over 75 then you should be a contender for 'world class' or 'leaders' league and your target scores for each foundation stone should be 0.5. Staying at the top is as challenging as trying to get there!

In our example shown in Figure 10.6, the target scores for eight foundation stones were set at the highest level (i.e. 0.5) whilst the targets for the remaining 12 foundation stones were set at 0.4. This is further illustrated in Table 12.1.

Table 12.1

	Foundation stone	Target score
1	Understanding the market place	5
2	Understanding the competition	5
3	Product and process innovation	4
4	Manufacturing planning and working with suppliers	5
5	Distribution management and working with customers	5
6	Supply-chain performance	4
7	Product safety	4
8	Industrial safety	4
9	Environment protection	5

10	Sourcing strategy	5
11	Appropriate technology	5
12	Flexible manufacturing	4
13	Reliable manufacturing	4
14	Manufacturing performance	5
15	Quality management	4
16	Financial management	4
17	Information technology and systems	4
18	Management skills and organization	4
19	Flexible working practices	4
20	Continuous learning	4

Normalizing performance

It is an essential requirement to normalize all measures when the benchmarking is done by comparing performance across companies. This is achieved by expressing all indicators and indices with common denominators. Established procedures are available (see Korlof and Ostblom, 1993) for normalizing non-comparable factors in an intercompany benchmarking project. The non-comparable factors can be due to:

- differences in operative content (e.g. order picking by cases or by full pallets);
- differences in scope of operation (e.g. maintenance costs with materials and parts or, labour only cost);
- differences in cost structure (e.g. operating costs including depreciation or excluding depreciation);
- National differences (e.g. postal services in sparsely populated Canada or, those in the densely populated Netherlands).

In the self-analysis exercise, the degree of non-comparability is likely to be less significant where the assessment is carried out for a company with a small range of product manufacturing in one country compared to a multi-product in a multinational company. For a multi-product multinational company the need for normalizing data should be critically reviewed. Very often the credibility of an exercise of this nature depends on how effectively performance measures can be expressed by a common base. Otherwise participants can be sceptical saying, 'Our business is so special?'

It is important to ensure that appropriate units and methods of measurement have been applied to the quantitative parameters mentioned for the 200 questions. For example, we have indicated in Question 135, that plant efficiency should be assessed as 'excellent' when operational efficiency (OE) is above 80 per cent. The definition of operational is given in Chapter 8. If the plant efficiency in your company is measured differently, you would obviously normalize your data accordingly.

Another issue related to 'normalization' is that we have given equal weighting to all 200 questions. Some team members of a company may believe that there should be different weighting for each pillar or foundation stone depending on the type or priority of a business. It is a logical consideration. However, as we have allocated a number of foundation stones – for example there are five foundation stones for the manufacturing facilities pillar – the sensitivity of differential weighting is not likely to be highly significant. Let us illustrate this point in Table 12.2.

Table 12.2

Pillars	Actual Score	Normal weighting	Adjusted weighting	Correction factor	Adjusted score
1. Marketing and innovation	13	15	10	0.67	9
2. Supply-chain management	13	15	15	1	13
3. Environment and safety	8	15	10	0.67	5
4. Manufacturing facilities	20	25	25	1	20
5. Procedures	8	15	20	1.33	11
6. People	<u>8</u>	<u>15</u>	<u>20</u>	<u>1.33</u>	<u>11</u>
Totals	70	100	100	6.00	69

By applying higher weightings to procedures and people, and lower weightings to marketing/innovation and to environment and safety, the manufacturing correctness factor changed from 70 to 69. The effect however is not significant. There is nevertheless some scope to adjust scores if the team applies extreme values of weighting (e.g. nil for environment and safety and 50 for manufacturing facilities). We believe that any manufacturing business must have all success factors or foundation stones, and thus extreme values of weighting should not occur.

Gap analysis tools

Gap Analysis is a systematic process, but in general it is basically 'organized common sense'. Therefore the approach and tools for gap analysis in total manufacturing solutions should also be simple but effective. Very often by using well publicized models, team members find that the models confuse the problem rather than clarify it. There is a danger of systematizing the problem rather than customizing the solution.

The practical tools are necessary both at the primary and secondary stages of analysis. The primary analysis involves a broad picture or 'a helicopter view' of the total business by identifying the overall performance gaps in each foundation stone. We discussed the tools for the primary analysis in Chapter 10.

Table 12.3 *Critical Examination. Source: Currie, 1964*

	Primary questions		Secondary questions
	Stage 1	Stage 2	
Purpose	What is achieved?	Is it necessary? If yes, why?	What else could be done?
Place	Where is it done?	Why there?	Where else could it be done?
Sequence	When is it done?	Why then?	When else could it be done?
Person	Who does it?	Why that person?	Who else could do it?
Means	How is it done?	Why that way?	How else could it be done?

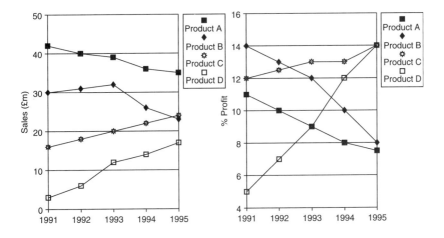

Figure 12.1 *Examples of line graphs in gap analysis*

The secondary analysis is concerned with the critical examination of the strengths and weaknesses identified by the primary analysis. Both the computer-aided and graphical tools of the primary analysis can be applied at the secondary stage, particularly to audit results and to carry out sensitivity analysis. In addition the team members can, depending on specific requirements, effectively apply:

220 TOTAL MANUFACTURING SOLUTIONS

- critical examination
- line graphs, bar charts, pie charts
- Pareto analysis
- 'total manufacturing process map'
- Z charts'
- cause and effect diagram

and similar analysis tools such as SPC tools.

Perhaps only 'total manufacturing process map' and 'Z charts' need some explanation as other tools in the above list are well known. However, for the sake of completeness we shall comment on these tools in the context of secondary gap analysis.

Critical examination is a powerful questioning tool of method study used by industrial engineers. It is not new. In fact, some managers and analysts have abandoned it because to them it is 'old hat' or 'déjà vu'. None the less, critical examination can be eminently practical to understand the reasons for performance gap in a specific area. The technique is simple and comprises primary questions (to ensure that the current process is clearly understood) and secondary questions (to investigate suitable alternatives to the present process). The questions – primary and secondary – query purpose, place, sequence, person and means (what, where, when, who and how). Table 12.3 shows an example of typical questions used in critical examination. The principles of this tool can be gainfully applied to non-value-added analysis and brain storming.

Line graphs are useful to describe a time series, such as comparing different companies' performances over time or for analysing the trends for relevant parameters of the same company. Figures 12.1a and 12.1b illustrate the trends of sales and profits for main products with reference to the secondary analysis of the results. Refer to Question 1 of our 200 questions (Chapter 9).

Bar charts can be gainfully used in comparing a specific performance measure between companies or between different units of the same company. Figure 12.2 shows an example of how the plant efficiency figures of different assets are compared with reference to the results for Question 135 of the 200 questions.

Pie charts can vividly illustrate the share of a total task, cost profile, profit margin, etc., by its various elements. Figures 12.3a-c show an analysis of a cost-effectiveness programme with regard to the secondary analysis of the results for Question 159.

Pareto analysis is a powerful tool for identifying A, B and C categories of products and activities so that appropriate priority and strategy can be applied. Figure 12.4 shows an example of Pareto analysis for stock holding with regard to Question 51.

There has been a considerable hype for promoting process mapping

GAP ANALYSIS 221

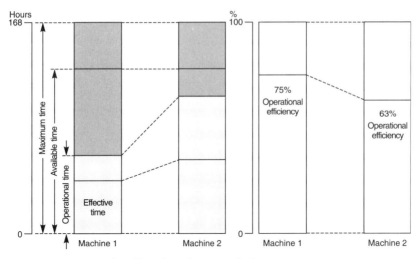

Figure 12.2 *Examples of bar charts in gap analysis*

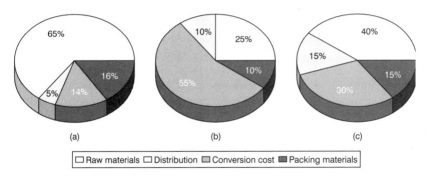

☐ Raw materials ☐ Distribution ▨ Conversion cost ■ Packing materials

Figure 12.3 *Examples of pie charts in gap analysis (a) Direct delivered cost. (b) Cost reduction efforts. (c) Savings achieved.*

by the proponents of both TQM and BPR. In practice, however, process mapping has proved to be both harmful and helpful. For an analyst it has proved very useful to describe all details of a process and thus help to understand the interrelationship of various operations and factors. On the other hand, to an outsider this offers little value and tends, with a large number of boxes and arrows, to confuse. As we indicated in Chapter 10 (see Figure 10.1) a simple flow diagram can be very useful to understand a business process and further details can be left to a systems analyst if so desired. A 'total manufacturing process map' is intended to be a simplified method of mapping to describe the various levels (from 'poor' to 'excellent') of each foundation stone.

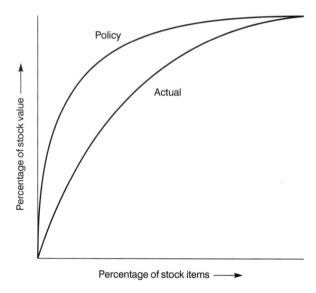

Figure 12.4 *An example of Pareto analysis*

During a gap analysis exercise this 'process map' can broadly locate where the business is at present and how far to go. It may also be expanded with a time scale as long as it does not complicate the purpose. Figure 12.5 gives an example of a 'total manufacturing process map' which shows the present position of the business used in the example shown in Table 10.1.

The Z chart is one of the benchmark analyst's favourite tools for comparing the performance of your own company against the benchmark of a competitor organization. This type of chart is fully described by Bogan and English (1994). Many companies have experienced that their strategic plans failed because they assessed a competitor's capability based on its current position. The Z chart quantifies the moving position. Thus a Z chart can be a powerful tool particularly for strategic planning. Figure 12.6 shows an example of a Z chart.

One of the most widely used 'SPC tools' in company-wide change programmes such as TQM is the 'cause and effect diagram'. This is also popularly known as a 'fishbone diagram' because of its obvious shape as shown in Figure 12.7. During a gap analysis exercise a 'cause and effect diagram' can be effectively used to map and analyse the root causes of a problem.

As we all know, the effectiveness of a tool or technique depends on the choice for application, the skill of the user, and the interpretation of results derived from its application. One should not carry a tool kit in search of problems, it should be the other way round.

GAP ANALYSIS

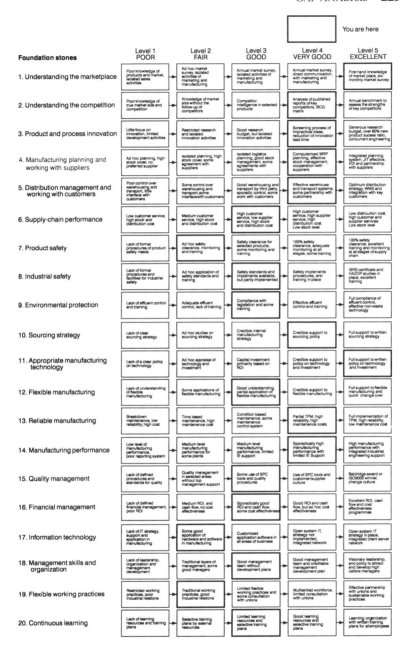

Figure 12.5 *Total manufacturing process map* (For enlarged version see page 270).

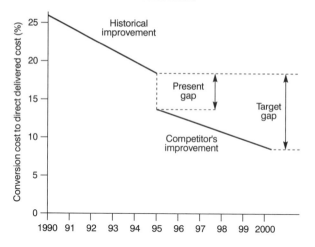

Figure 12.6 *Example of a Z chart*

Lessons learnt

The quantitative analysis of the performance gap is only part of the gap analysis. An important by-product of the analysis is that it makes the team delve into the details of the business processes. This enables the team to identify and understand the underlying causes that explain why the performance gap exists. Although the main purpose of the gap analysis is the identification of opportunities for improvement, it also offers a number of other benefits including:

- An understanding of the total manufacturing business process and the success factors underpinning world class performance and best practices.
- Bringing together the business perspectives of a multi-functional team comprising marketing, technical, financial, logistics and human resources.
- In-depth knowledge of foundation stones by specialists in the team, whether they are related to performance or practices.
- A general improvement in performance and practices by focused attentions during the analysis even before the start of the implementation plan.

Summary

In this chapter we have shown how gap analysis is a powerful tool for

GAP ANALYSIS 225

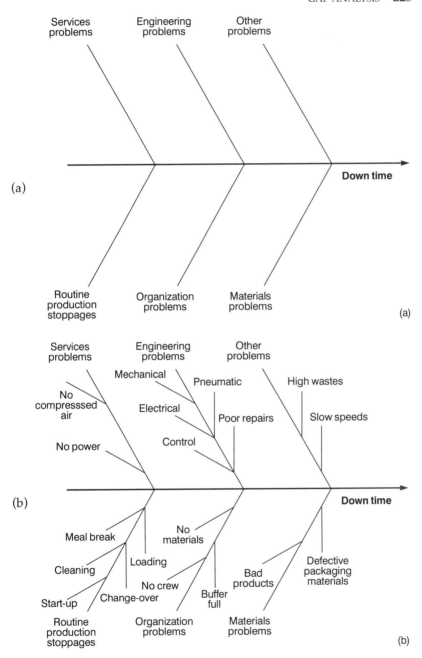

Figure 12.7 *Cause and effect diagram: downtime in a packing line*

identifying where action has to be taken for a company to achieve world class status. We demonstrate how systematic analysis and accuracy of measurement will show the way. Various techniques for analysis are considered. We point out that a significant by-product of gap analysis includes identification and delving deeper to understand the root causes of problems. The very nature of this delving for information, by a team drawn from various key functions, will draw together functions and highlight commonality of purpose and the need for teamwork throughout the organization. Identifying the problem and the root causes is the objective but the understanding and agreement of problems and actions by an influential group of company-wide people will in itself strengthen the company and bring forth untold advantages.

References

Bogan. C.E. and English, M.J. (1994) *Benchmarking for Best Practices*. McGraw-Hill.

Currie, R.M. (1964) *Work Study*. BIM Publications.

Karlof, B. and Ostblom, S. (1993) *Benchmarking*. John Wiley & Sons.

13

Improvement strategy

Probable impossibilities are to be preferred to improbable possibilities.
Aristotle

Once we have established gaps in our performance the natural reaction is to take immediate, almost urgent action to correct the situation. However, we caution patience. Having come this far it is important that we do not take precipitate action. Our philosophy is that there is no such thing as a quick fix. The best results are achieved by careful consideration of all factors.

A vital stage between gap analysis and the implementation phase is the selection of the appropriate improvement strategy. Following the results of the self-analysis and performance gap analysis the team should develop a recommendation for change. A lack of in-depth investigation of what is appropriate for a company has caused many sound change programmes such as TQM and BPR to fail. Recent publications are quite openly, but wrongly, criticizing well proven concepts as 'management fads'. We liken what has often happened to the application of wonder drugs by a doctor without the diagnosis of the illness, and without taking into account the condition of the patient, with the 'cure' being more deadly than the ailment. M. Lynn in the *Sunday Times* (1995) identified the danger of 'management fashions' as 'They can be treated as instant panaceas, and that is a mistake'. In another article in *Fortune* magazine Collins (1995) has gone further by declaring, 'Re-engineering and other prevailing management fads that urge dramatic change and fundamental transformation on all fronts are not only wrong, they are dangerous'. Even CSC Index (the firm pioneering BPR) from a survey in 1991 reported that 25 per cent of about 300 north American companies involved in BPR are not meeting their goals. Other BPR experts estimate an astounding failure rate of 70 per cent. This does not mean that re-engineering should not be considered; it means that the diagnosis has to be sound and that the choice and applications must fit the condition of the patient company.

In short it is nothing less than a sound strategy to review fundamental criteria before adopting an improvement programme. We suggest that the three key criteria are improvement category (as described below), core strengths and the present position.

Improvement category

The approaches and characteristics of an improvement programme depend on both the degree of change (i.e. technical or strategic) and the pace of change (i.e. fast or longer term). Figure 13.1 illustrates the four broad categories of performance improvement. This type of conceptual mapping has been widely used by consultants (e.g. Ernst & Young, and Bogan & English). The categories are focused improvement (fast tactical change), continuous improvement (longer-term tactical change), focused restructuring (fast strategic change) and process re-engineering (longer-term strategic change).

		Pace of change	
		Fast	Longer term
Degree of change	Tactical	Focused improvement	Continuous improvement
	Strategic	Focused restructuring	Process re-engineering

Figure 13.1 *Improvement category*

A tactical change usually involves small-scale incremental improvements. A strategic change represents improvements in much larger steps and has a deeper time horizon.

Focused improvement

This type of improvement may not appear to be radical but it is achieved within a short time scale and very often without any significant capital expenditure. Such changes are also relatively easy to implement and therefore form a big part of a company's cost effectiveness programme.

Continuous improvement

The application of total quality, flexible working practices and total productive maintenance are examples of the continuous improvement category. These types of programmes are invariably company wide, requiring cultural change over a long period.

Focused restructuring

Focused restructuring often aims at combining activities and departments. Such major organizational changes resulting from an acquisition or site closure may require an immediate down-sizing or structural re-organization. Focused restructuring often aims at combining activities and departments.

Process re-engineering

Business process re-engineering results in strategic changes with a deep time horizon and is achieved by redesigning the core processes of a business. The new operating model is often influenced by the success of processes adopted by benchmarking partners.

The choice of a particular improvement approach should depend on a company's specific requirements and commitments. It is also likely that a company may apply a carefully selected portfolio of all improvement approaches. However the key message is that the selection should be based upon your own self-analysis and not by the pressure of fashion or fads.

The company's core strengths

As we discussed in Chapter 11, a company must identify its core strengths and incorporate them in its mission statement. Collins (1995) emphasizes that leading-edge companies have successfully adapted to a changing world without losing their core values. They have understood the difference between fundamental principles and daily practices. For example, according to Collins, Disney has diversified its product strategy from cartoons, to feature films, to theme parks (Disney Land), but has maintained its central ideology (vision), of bringing happiness to people.

It is important to identify a limited number of core principles (usually not more than five) and any improvement programme or change process must not deviate from these fundamental ideals. A core ideology does not arise from the pursuit of competitive advantage. Core values must be distinguished from, and not confused with, business practices or 'sacred cows'. Once core principles are clearly understood, and encapsulated in the mission statement, then, and only then, is it safe to consider changes.

Furthermore, a company's core strengths need to relate to its product and marketing strategy. A change programme must take into account the strategic measures in pursuit of exploiting a dominant position. For example a ruthless cost-effectiveness programme is a sensible approach in mature markets. Similarly, knowledge gained in developing markets may significantly benefit mature markets. Improvement programmes related to quality logically should be applicable to products with a shorter life cycle, whilst those related to performance should be more appropriate for established products.

An aid to the selection of an improvement strategy (based on the criterion of a company's core strengths) should be derived from the understanding of its mission and objectives rather than from the self analysis of the 200 questions. In short, when you understand your own mission and objectives you understand your core strengths. The 200 questions identify areas where corrective action is required. But your strategy for improvement must be based on your prevailing culture and ideals.

The company's present position

Following a self-analysis with the 200 questions, and a further review of the performance versus practice matrix, a company can locate its position within one of the five categories, namely leaders, behaviourists, structuralists, plodders and laggers (see Figure 10.7 in Chapter 10). The choice of an improvement strategy can be significantly influenced by your identity with one of these five categories. Let us examine how this approach can be applied.

Leaders

This category of companies has achieved world class performance and global market leadership. They can sustain their advantages by continuing the application of best practices. For these companies there may not be any road map to follow, as they themselves are creating it. Nevertheless, they cannot afford any form of complacency, or any slippage, as the pressure to stay at the premier league is decidedly tough. In any event our philosophy is that perfection is never attained, there is always room for improvement. Therefore, in addition to maintaining a continuous improvement culture and retaining their core strengths, their improvement strategy should contain:

- further expansion and penetration capability with cost-effective products in the global market;
- continuous investment in improving relationships with stakeholders (e.g. investors, suppliers, customers and employees);
- a relentless pursuit to innovative new products and new business concepts.

Behaviourists

Companies in this category will have invested heavily in TQM initiatives and have a best practice culture. The systems (including IT), procedures and infrastructure for change programmes are in place. Employee involvement and team work are actively encouraged by management. These are continuous learning companies but are yet to reap the benefits of their efforts. There are two groups in this category. For the group with a small performance gap, because the infrastructure of best practice programmes and change culture is already in place, they should find the route to world-class status reasonably straightforward.

On the other hand the behaviourists, with a large performance gap, will have to face the reality that the rest of the journey is going to require discipline and a concerted company-wide effort. The momentum of TQM initiatives may have been lost and perhaps members of the company might be disillusioned.

To re-energize and to re-kindle enthusiasm for continuous improvement programmes the improvement strategy for behaviourists will need to include:

- a market-led manufacturing strategy towards reducing operating costs and improving customer service;
- programmes focused on performance improvement and cost effectiveness;
- abandonment of change programmes which are not delivering results and which are seen as fads by many employees.

Structuralists

The structuralist group of companies are profitable and usually have a dominant share of local and regional markets. They may be ahead of the game in their own patch but in fact may well be vulnerable to emerging global competition. Their business strategy has been to emphasize targets and results and to date this has been successful, but in general they have neglected to develop a change culture and a philosophy of best practice. It is likely that without adopting a best practice culture, and without the appropriate people and systems infrastructure, they may not go the distance to join the leaders' league.

This group first has to recognize that action and change are needed and then to adopt an improvement strategy. The strategy should include:

- carefully designed continuous improvement programmes (including TPM) to enhance change culture and continuous learning;
- programmes for meeting product safety, and industrial safety environmental protection;
- additional emphasis on longer-term plans for improving human resources issues including flexible working practices.

Plodders

The companies in this group may possess pockets of excellence in both performance and practices, but their overall strengths are not good enough to compete in a global market. A high proportion of companies fall into this category. Their local strategies have shown good results in specific areas but they lack a coherent and potent leadership to benefit from the best practices of each unit. We have found that it is not uncommon to observe world class manufacturing lines alongside less efficient lines in the same manufacturing site. Another example is a manufacturing organization that has its own high street retailers. This organization has an admirable policy of seven-day stock turn in the retail outlets but the balance sheet showed that they held eight months of finished goods in their distribution warehouses, and a further seven months of raw materials at the factories.

Plodders will require radical changes in both performance and practices to progress towards the leaders' league.

In addition to retaining the pockets of excellence the improvement strategy for 'plodders' should include:

- the recognition and removal of inhibitors so that their own people skills and best practices can be developed to their full potential;
- retention of the core values of the company, and a selective but aggressive application of business process re-engineering in the critical areas of the business;
- encouragement of networking and promotion of best practices within the organization.

Laggers

This is the lowest scoring group in both performance and practices. It is unlikely that these companies can become world class in the foreseeable future. It is however possible for a company to operate in a protected market without long-term best practice or high performance. But as local protection is on the wane, these companies must make drastic changes for their survival. There is no merit for rushing into automation or TQM. Many companies in this group will be working with limited human and financial resources. Changing business procedures will have limited impact as these will only speed up poor processes.

In addition to holding onto the strengths, if any, in selected products, the improvement strategy for 'laggers' should be focused on short-term changes including:

- aggressive restructuring, down sizing and asset utilization programmes;
- systematic simplification of manufacturing and business processes (namely, industrial engineering) toward high performance and quality with minimum investment;
- continuous learning and communication programmes to develop the company's human resources and also to sustain the morale of the workforce;
- practical longer-term plans for both continuing business and possible mergers with more profitable organizations.

Summary

Recently much attention has been focused on the danger of management fads reinforcing scepticism and resistance to change amongst manufacturing managers. There are also doubts about endangering the core values of a business by following fashionable, but inappropriate, business practices.

In this chapter we have identified the types of improvement processes and provided a systematic approach of adopting an improvement strategy that is most likely to be appropriate for the particular business. Naturally, the categorization of a manufacturing business and the selection of a strategy is not always clear cut. In the real world a perfect model does not exist and there will always be a 'grey area' of overlap. Nonetheless, our methodology should provide useful guidelines for enhancing the proportion of objective analysis, thus helping the selection of improvement programmes.

References

Collins, J.C. (1995) Change Is Good – But First Know What Should Never Change. *Fortune*, May 29.

Lynn, M. (1995) Book Sales Rigging Claim Shakes Industry. *Sunday Times*, 13 August.

14

Implementation plan

> *Each morning sees some task begin,*
> *Each evening sees it close;*
> *Something attempted, something done,*
> *Has earned a night's repose.*
> H.W. Longfellow

Many organizations start the improvement programme without going through the earlier stages of identifying the real requirements. This is usually the result of accepting the latest management fads without thinking through all the steps described in previous chapters. Change programmes that are not carefully planned and managed are doomed to failure, because not only is it likely that the improvement strategy will be wrong but also the necessary commitment and culture will not have been developed.

Many benchmarking projects have been well carried out and produced excellent reports and action plans, but due to poor implementation planning and a real will to make things happen little has been achieved. The knowledge gained through benchmarking of your own strengths and weakness can only lead to disillusionment of those involved, if no improvements are made or attempted. Disillusionment among a few key staff will be contagious and will result in a lowering of morale which will spread throughout the organization. Thus a benchmarking exercise, no matter how well carried out, if no follow up action is taken, will lead to negative results.

Therefore both benchmarking and implementation should be considered as one, and not as separate initiatives. Both are phases of a common continuous improvement process. As Bogan and English (1994) say, 'Segregating benchmarking planning and implementation planning is like separating Siamese twins who share vital organs'.

Making changes and improvements should be a continuous process, but to sustain continuous change is as difficult as initiating and implementing change. To keep the momentum going, it is necessary to evaluate if the change process has produced results and to keep developing ongoing improvement activities.

The success of any project is underpinned by management commitment, organization and resources. Building a commitment for all the stakeholders, inside and outside the company, involves the understanding of why improvement is needed and the nature of improvement. It is a common phenomenon for various factions to appreciate why a change is required but at the same time to believe that the need to change does not necessarily apply to them. As we have said in earlier chapters, the culture of the organization has to be such that everyone from the cleaner to the chief executive believes that they have a personal part to play in making changes. The prerequisite for change is the vision and the will to change based on a culture that will accept change. As top executives of Motorola are reported to have said, a thriving company constantly transforms itself while adhering to beliefs that are not subject to change.

It is vital that detailed discussion and agreement occur throughout the company as to what, how, when and where change should take place and who should be involved. We know of many process maps and implementation steps for improvement programmes. Motorola has five steps, Xerox has 12 steps and the Strategic Planning Institutes' Council for Benchmarking has a five-step model (launch, organize, reach out, assimilate and act). All of these models are tried and proven methods of implementation, and although each has a different approach they each include a structured approach. We favour a four-stage process as illustrated in Figure 14.1. This map covers the total improvement process. Our map is provided as guide only on how to organize, and how to make an improvement process happen.

Start-up

The key task for senior management is to decide what improvement opportunity areas have the greatest impact for the business. However a significant number of companies that initiate a change programme do so because either they feel threatened for survival or they have become victims of a three lettered management fad. Our recommendation is, of course, before any improvement is attempted, that self-analysis to identify the weaknesses and the gaps in performance takes place. A self-analysis process does not start on its own. Any benchmarking programme, especially the one described in this book, requires full commitment, preparation and organization. The start-up phase contains three major milestones:

IMPLEMENTATION PLAN 237

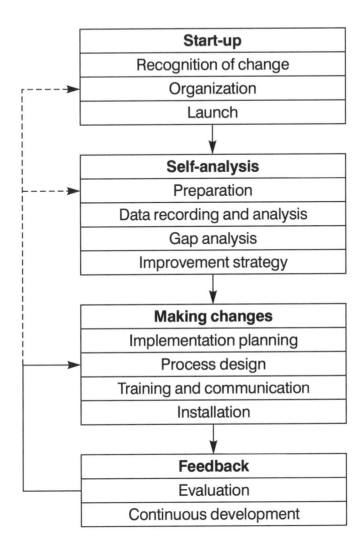

Figure 14.1 *Total manufacturing implementation process*

- Recognition of change
- Organization
- Launch.

Recognition of need for change

It is vital that top management and the board wholeheartedly recognize the need for a change programme. This recognition may be prompted by a reaction to current company performance, threat from a new competitor, or a strategic change (e.g. merger or an internal report from any of the key stakeholders). The board and management must believe that serious action has to be taken.

The least risk approach is to conduct your own benchmarking in the systematic way we have detailed in the previous chapters. Japanese success has clearly demonstrated that the 'longer way around is the quickest way home'. Take one step at a time and be patient; make big improvements through a series of small improvements each being non-disruptive and 'right first time'. Major, panic-driven changes can destroy a company. Poorly planned change is worse than no change.

At this stage it may be helpful to conduct a limited number of consultation workshops with key stakeholders to acquire agreement and understanding about the need to change.

The outcome of this milestone is the full commitment of top management to go ahead with the programme with the support of the stakeholders. The programme begins with the formation of a project team.

Organization

The organization phase involves a clear project brief, appointment of a project team and a project plan. The project brief must clearly state the purpose of the project and the deliverables expected from the project.

There is no rigid model for the structure of the project team. The basic elements of our recommended project structure are shown in Figure 14.2.

Steering committee

To ensure a high level of commitment and ownership to the project, the steering committee should be drawn from members of the board and include senior management. Their role is to provide support and resources, define the scope of the project consistent with corporate goals, set priorities and consider and approve the project team's recommendations.

IMPLEMENTATION PLAN 239

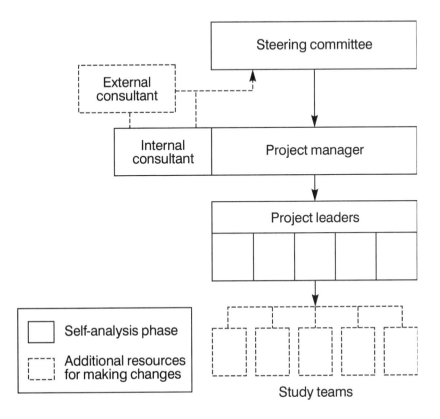

Figure 14.2 *Total manufacturing project organization*

Project manager

The project manager, or more correctly the project leader, should be a person of high stature in the company, probably a senior manager, with broad experience in all aspects of the business and with good communication skills. He or she is the focal point of the project and also the main communication link between the steering committee and the project team members.

The project leader's role can be likened to that of a consultant. However if a line manager is given the task of project leadership as an additional responsibility to their normal job, then an experienced staff manager can be co-opted to support the leader. The role of the leader is to a great extent similar to Hammer and Champy's 'csar' in *Reengineering the Corporation* (1993). The project leader's role is to:

- provide necessary awareness and training for the project team, especially regarding multi-functional issues,
- facilitate work of various project groups and help them develop design changes, and
- interface with other departments and plants.

In addition to the careful selection of the project leader, two other factors are important in forming the team. First, the membership size should be kept within manageable limits. Second, the members should bring with them not only analytical skills but also in-depth knowledge of the total business covering marketing, financial, logistics, technical and human resources. The minimum number of team members should be three, and the maximum number should be six. Any more than six can lead to difficulties in arranging meetings, communicating and in keeping to deadlines. The dynamics within a group of more than six people allows a pecking order to develop and for sub-groups to develop. The team should function as an action group, rather than as a committee that deliberates and makes decisions. Their role is to

- provide objective input in the areas of their expertise during the self analysis phase, and
- to lead activities during the making changes phase

For the project leader, the stages of the project include:

- education of all the people of the company,
- gathering the data,
- analysis of the data,
- establishing study teams to recommend changes,
- regular reporting to the steering committee, and
- regular reporting of progress companywide to all the people of the company.

Obviously the project leaders cannot do all this work themselves. But they have to be the sort of people who know how to make things happen and who can motivate people to help make things happen. To assist in various phases study teams should be formed to work with the project leader.

We strongly recommend Eddie Obeng's book *All Change! The Project Leader's Secret Handbook* (1994) as essential reading for project leaders.

Study teams

In general, study teams should be formed after the self-analysis phase, but in selected areas. However some members of study teams can assist in the original data collection phase and also in the analysis phase. The members of the study team represent all levels across the organization and are the key agents for making changes. Their role is to develop design changes and submit recommendations to the Project Leader.

External consultants

Use of an external consultant at various stages of the project might be useful to supplement your own resources. However a consultant cannot know your company as well as your own people do. It could be argued that a consultant will not only bring his or her expertise and experience but will also act as a catalyst during the total implementation process. Likewise, in the initial stages, consultants can be effectively used in training the people of your company in both analytical tools and in assisting with culture change. In our opinion the best time to employ a consultant is probably after your self-analysis has been completed and after the selection of a specific improvement strategy has been made. We recommend the employment of a consultant who is a specialist in a particular field. For example if the problem is with logistics then a consultant in that field could be employed. We do not favour using external consultants as project leaders.

Launch

It is critical that all stakeholders (e.g. managers, employees, unions, suppliers and key customers) who may be immediately impacted by the programme are clearly identified. Internal stakeholders must be consulted and kept fully informed at every stage of the programme. After the organization phase the next milestone is the formal launching of the programme. The nature of launching can be either low key or a big bang. Our recommendation is that before the self-analysis a low-key, but definitely not a secretive, approach is more appropriate. Too much excitement and too high an expectation could be counterproductive if it leads to uncertainty. It will be only after self-analysis that an improvement strategy can be finalized. A high profile launch would therefore be more appropriate once the strategy has been approved by the steering committee.

The nature of the launch sets the tone for how future communication will take place and identifies the ownership of the project. It is absolutely essential that strong and visible support be given by senior management.

Self-analysis

After the start-up, the project team will be involved with the self-analysis stage of the project. This is basically do-it-yourself benchmarking. It is worth repeating that it is vital to assess your requirements before you start to implement any improvement programme.

We have described in detail in Chapters 10, 12 and 13 how team members can carry out a self-analysis and the basis on which they can recommend a suitable improvement strategy. The main steps and activities involved during this stage can be summarized as:

1. Preparation

- Collect information about your own company based upon the checklist given in Chapter 10.
- Analyse and understand your own business and manufacturing processes.
- Study and understand the six pillars and 20 foundation stones as described in Chapters 3 to 8.
- Obtain published reports on the best practices of successful companies.
- Study and understand the 200 questions described in Chapter 9.

2. Data recording and analysis (see Chapter 10)

- Carry out the assessment and scoring process for the 200 questions with regard to the best understanding of your own operations.
- Calculate your manufacturing correctness factor.
- Conduct a primary analysis of the results with the aid of graphical presentations including our rating profile, the 'spider diagram'.
- Conduct a performance versus practices analysis and identify the category of your organization (leaders, behaviourists, structuralists, plodders or laggers).

3. Gap analysis

- Identify the core values, priorities and mission of your company (see Chapter 11).
- Re-assess your target scores based upon your mission and core values.
- Re-examine the scores obtained during the primary analysis of the

results given for the 200 questions (see Chapter 12).
- Identify your strengths/weaknesses and the gap between your target and actual performance as described in Chapter 12.

4. Improvement strategy

- Prepare a report of your key findings resulting from the self-analysis stage of the study.
- Include in the report recommendations for improvement programmes as described in Chapter 13.

Making changes

In this phase the change process moves on to the action programme to make the changes happen. Having chosen the improvement strategy, the detailed work of implementing the changes will be influenced by the strategy. The strategy may be a confirmation of well-known improvement programmes (e.g. TPM for the factories and JIT for supply-chain management). We have found it to be effective to name the total initiative (e.g. Project 'UTOPIA') so that everyone in the organization can identify it as a 'single issue improvement culture'.

There is an abundance of publications regarding change management and company-wide implementation programmes. In addition, how change is tackled will vary according to improvement strategy, as we have stated earlier. Therefore we shall outline key milestones of implementation planning, process design, learning and installation which are likely to be relevant for all programmes for making changes.

Your own benchmarking or self-analysis helps your company find its way through two aspects of change. First, it provides a road map describing areas which require more changes than others, second, the self-analysis phase of the programme provides insight and experience on how to establish a single culture that transcends divisional boundaries.

Implementation planning

It is possible that after spending several weeks with the self-analysis phase employees outside the core project group may demonstrate scepticism. If this shows signs of occurring it may be necessary for top management to relaunch the initiative, for example 'Utopia Stage Two'.

If this is done the project team will need reconfirmation. There is an obvious advantage of continuing with some of the same people involved in the self-analysis stage. Their experience, new-found company-wide knowledge, and their belief in the recommendations should not be under valued. The number of those now directly involved will increase by the formation of study teams for the areas selected for study, and as discussed earlier specialist external consultants might now also be introduced.

The project leader will be responsible for writing implementation plans indicating key tasks, responsibilities, deliverables, resource requirements and target dates. It is recommended that the project plan will include a critical path, and that periodic reviews and reports be made by the project leader to the steering committee.

Process design

Process design relates to the actual transformation of an operation, procedure, organization or facilities from the current state to a desirable future state.

The nature of process design depends on the category of improvement as illustrated in Table 14.1.

Table 14.1

Improvement category	Programme	Examples of process design
Focused improvement	Labour productivity	Improved method and layout of order picking
Focused restructuring	Site development house	Relocation of boiler
Continuous improvement	Total productive maintenance	Lubrication and cleaning of machines by operators
Focused re-engineering	Supply-chain re-engineering	Direct delivery to urban retailers

The design and redesign of a process by a study team must incorporate the ideas of the people who will actually be affected by the change. The tools of gap analysis can be systematically applied to develop and evaluate a process design. Innovation in process design will result from the readiness to abandon traditional thinking and to be able to imagine a 'green-field' approach to a process.

Some design processes (usually focused improvement) are accomplished at the project leader level and others (usually strategic changes, and changes requiring capital expenditure) are approved through the steering committee. However the steering committee should be kept informed of all changes and their progress. There is no advantage to be gained by being secretive. A world class organization does not have secrets or hidden agendas.

Training and communication

We have emphasized in Chapters 7 and 8, drawn from the experience of several companies, that training and communication are the life blood of a change process. It is so fundamental, so obvious, that we feel we insult your intelligence by mentioning the need to keep everyone in the company fully informed as to what is happening. At the risk of being pedantic we will however list some of the key communication benefits and needs.

1. The objective is to share information and change processes among the stakeholders at all levels of the organization, e.g.:

- Top management, and the board, must understand enough about the improvement programmes to know how the changes will affect the business. They must be able to know what is happening and to show leadership so that things will happen.
- Study team education. The study team needs to have a detailed understanding of what is planned for their area and a good overall understanding of the big picture. They are the ones who will be responsible for working with and training people in process design changes in sections and units of the company.
- Middle management and staff education. While everyone cannot be on the study teams, everyone has a role to play in the improvement programme. Therefore everyone on the staff must be informed of how their work will be affected.
- Employee training. No change process will work if the employees on the shop floor oppose it either directly or indirectly. Employee involvement and training are vital for an implementation plan. Many of the ideas for improvement in a programme such as TPM come from the operators.
- Communication to unions. It is critical that the representatives of unions and other staff representative bodies are kept informed at critical stages of the implementation of how the change process will affect their members.

2. The communication among the stakeholders should be full and open. A change progamme cannot be built upon any false pretence. Success depends on trust.

3. Although the study team may be responsible for making changes at the factory at floor level the employee learning programme should be properly structured.

- There should be a learning manager with a focused role.
- On-the-job learning should be accomplished through line supervisors.
- An external human resources consultant may be valuable to guide the learning and to effect a culture change.

Installation

The installation phase involves the planning and physical actions necessary for putting the improved process into place. The project manager may co-opt industrial engineers to facilitate engineering changes. Separate capital proposals may be required, and these should be channelled through the steering committee. Other likely expenditure will include modification of equipment and for moving equipment. It is important that proper authorization is obtained for any expenditure before it is incurred. The project has to lead by example and cannot be seen to be taking short cuts.

The installation stage consists of a large number of concurrent and parallel activities including selection of equipment, revising layout, improvement of process capability, commissioning, training and so on. It is useful to prepare a project schedule showing the critical path and all the necessary resources. For very large projects, the expertise of external engineering consultants may be sought to carry out a 'front end' engineering study (e.g. feasibility study, conceptual engineering and preliminary engineering).

When a significant change in process design is involved, the installation should start with a pilot programme in a selected line or plant so that the process is proven and operations are understood by the employees. The project in the pilot area may run for several months as a learning centre. Some people understand a system conceptually but cannot accept it unless they can see it in action. Pilot projects can demonstrate results and validate the purpose of the change. It can be a great advantage to move along the learning curve by installing a pilot line ahead of a completely reorganized or new plant.

Another important issue of installation, unless the project is on a green-field site, is that it should be planned to create a minimum disruption to current operations. When a disruption is unavoidable, a stock build up to cover a limited period of installation may be necessary to continue to meet the market demand. Factory house keeping and safety standards should not be compromised during a change. Once again we repeat the project must lead by example, it cannot be seen to be condoning short cuts or sloppy practices.

For some people (most of us) details are tedious and it may be difficult to maintain the momentum of earlier design efforts throughout the company. It is therefore important to include two actions in the installation programme:

- Installation should be a logical extension of earlier activities for those people who were involved in the process design stage. They will believe and be committed to their recommendations.
- Installation activities should be continuously checked against target results and time frame.

Feedback

A change programme in total manufacturing solutions and the achievement of world-class status is a never-ending journey toward continuous improvement. The phase of feedback involves the continuous need to sustain what has been achieved and to identify further opportunities for improvement.

It is at least as difficult to sustain changes as it is to design and install them. Keeping the change process going by regular feedback is a different process from that of making changes. It usually calls for different approaches and sometimes the responsibility of this phase may shift to a different team.

The feedback phase contains two interrelated milestones – evaluation and continuous development.

The progress of the changes should be monitored at regular intervals, usually by comparing the actual results with target performance levels. The audit team may decide to carry out a sample self-analysis by selecting from the 200 questions.

The frequency of sample analysis may be every year, in addition to the continuous monitoring of key performance indicators along the way.

Ongoing development

The evaluation or audit may reveal that some changes are working well and some changes will need re-designing. There is a need to continuously align and adjust the performance level with specific needs.

A change programme can attain its full effect only by repetition. When you have been through the self-analysis for the first time you not only have a working model that can be used again, you also have a proven project organization experience to carry the changes through to their implementation.

You are likely to have a target moving progressively from internal, to external, to functional, to best practice (see Figure 14.3). External targets refer to comparisons with competitors. Functional targets refer to comparisons with similar functions without regard to the type of industry in which they operate. Thus functional targets offer the greatest potential for funding major improvements. Our methodology should assist you, and external consultants if employed, to achieve both external and functional benchmarking.

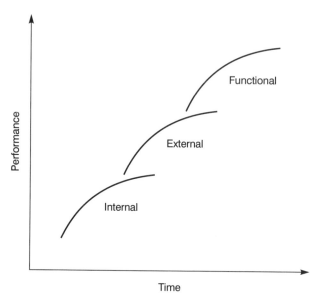

Figure 14.3 *Total manufacturing solutions: moving targets for excellence*

Time scale

The time scale of total change process from start up to feedback is expected to last for several months and naturally the duration is variable. The time not only depends on the degree of gaps in different foundation stones but also on risks involved and resources required. Four factors of the business can favourably affect the time scale:

- Full commitment of top management
- Good cash flow of the company
- The workforce is receptive to change
- A number of products retain a competitive niche in the marketplace

Given the existence of favourable factors, typical time scales for a change programme are likely to be as follows:

IMPLEMENTATION PLAN

Start-up	2 to 6 months
Self-analysis	3 to 9 months
Making changes	15 to 24 months
Feedback	4 to 24 months (plus ongoing)
Total	**24 to 63 months**

Figure 14.4 shows a hypothetical summary schedule for implementing a total manufacturing change process in a medium-sized company. The diagram depicts an order of magnitude only. In addition, the exact sequence of activity would vary according to the company undertaking the programme. The map is not linear either, it may need frequent looping back to reflect the continuous learning with ongoing progress and development. However, there is no doubt that there is no short cut route to a total improvement process.

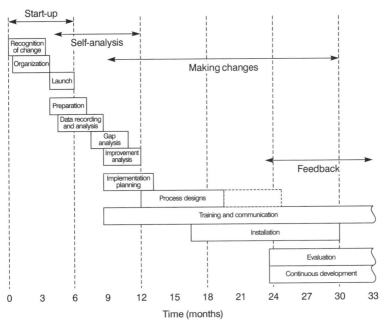

Figure 14.4 *Total manufacturing solutions: a hypothetical example of a timetable*

Conclusion

This chapter discusses how to make it all happen. Many a benchmarking exercise has produced excellent reports, many plans have been well researched and look good on paper. But until something actually happens, until some results are seen, all that has really occurred is expenditure in time and money. Our objective is not to waste your time and money, our objective is to get you a worthwhile pay back for your efforts.

Our methodology will enable you to determine your strengths and weaknesses and the gaps in your manufacturing correctness factor. But our method calls for a systematic step-by-step approach. Short cuts are always appealing. Believe us, our method is the short cut, there are no further short cuts available.

Our implementation process begins with the start-up phase. This includes the board and the senior management first recognizing that change is essential and being committed to making changes happen. We then discuss the organization needed for implementing the change programme, including the establishment of a high-powered steering committee, appointment of a project leader and a project team drawn from all the key functions of the organization.

The project team will need to be kept small, it is an action group not a planning committee. The team will need careful briefing and will need to fully study this book. The team has the important role of carrying out the self-analysis and benchmarking of the company. To do so they will need to be able to understand each of the 200 questions, and also be able to dispassionately rate the company against each of the questions. Gathering of data for their use might at first sight seem to be a onerous task, but most of the information should already exist and on further examination should be found to be readily available. We discuss the help that the team will need to get the information they require, including assistance from people who will later form the nucleus of the study teams.

Once the gap analysis has been carried out, the next stage will be to develop an improvement strategy. At all stages the steering committee will need to be regularly advised by the project leader of the progress of the team. The strategy will come from recommendations of the project team, but will be determined by the steering committee.

At all stages of the project, it is essential that not only is the steering committee kept informed, but that all members of the organization right down to the lowest levels are kept appraised of the aims and activities of the project. Initially, while the data gathering and the analysis of the 200 questions are being carried out, many people might not be aware that much is happening. It may therefore be necessary to give the project a high-profile relaunch once the strategy has been determined.

All the people of the organization have to understand the purpose of the project, believe in it and wholeheartedly support it. It goes without saying that the lead must come from the top of the organization. To get the full commitment of the whole organization, a major change in culture may be needed. Changing the culture is likely to be part of the improvement strategy.

In Figure 14.4 we give a hypothetical time table for total manufacturing solutions to be implemented. This time table shows that almost three years will be needed. This might seem a long time, but for some organizations the changes could take over five years to internalize. But even then you are not finished. The very essence of a quality programme designed for an organization to reach world-class status is that there is no end to the progress. Change is continuous and inevitable.

References

Bogan, C.E. and English, M.J. (1994) *Benchmarking for Best Practices.* McGraw-Hill.
Hammer, M. and Champy, J. (1993) *Reengineering the Corporation.* Harper Business.
Obeng, E. (1994) *All Change! The Project Leader's Secret Handbook.* Pitman.

15

Final analysis: 'the big picture'

Open the second shutter so that more light can come in.
　　　　　　　　　　　　　　　Johann Wolfgang Goethe

In Chapter 1 we said; 'in a global marketplace manufacturing is the main competitive weapon, and real wealth can only come from the physical adding of value in the manufacture of tangible products'. There are several important issues raised in this 30 word statement, namely:

- We are all now competing in a global marketplace. Make no mistake about it. There is nowhere to hide; national boundaries and governments no longer provide protection against overseas competitors.
- Manufacturing, not service, is the key competitive weapon. But, as you have progressed through this book, you will have realized that it is not possible to separate service from manufacturing. We cannot think of any manufacturing organization that is not also competing in the level of service that it provides. In a global marketplace first-class service, in the sense of delivering the right product, in the right quantities, at the right time, and with good post-delivery follow up, is taken for granted.
- Real wealth for a nation can only come from the physical adding of value in the manufacture of tangible products, the key words being 'adding value'. Any activity and any input that do not add value in the creation of tangible products add to the cost. We have to be ruthless in the way we look at all costs. Obviously some non-value-adding activities, such as transport and repairs related to manufacturing, will necessarily occur. Your aim should be to establish what is adding value and what is not adding value, and what is necessary and what is not. If an activity is not adding value and is not deemed to be necessary, then, if you are to be competitive, it must be minimized.

This book shows how efficiency in manufacture and service can be achieved – and how eventually a company can become a world class manufacturer by attaining excellence in both performance and practices. Achieving world class status will not be easy and requires total, not piecemeal, manufacturing solutions.

Our six pillars and their 20 foundation stones provide the structure for total manufacturing solutions. The importance of each pillar, and their foundation stones, is explained in Chapters 3 to 8.

This book is not a fad, it is not a three-letter acronym. We are not peddling a new panacea which we try to fit to all situations. We recognize that all companies are different and what will work for one organization will not necessarily be appropriate for another. We don't give the answers, but we show how you can identify your own strengths and weaknesses and how, from an analysis of these strengths and weaknesses, you can determine what has to be done. It is not enough for a company to know that they are losing market share, or that they have a high turnover of good technical staff. Such facts are usually self-evident.

Our belief is that manifestations of obvious problems will only be symptoms of a deeper malady. In the search for a healthy company we stress that all six pillars have to be looked at. Even if one pillar is performing reasonably well, this might only be in comparison to the other pillars. It would be a mistake to take a short cut and to try and get away with rating say only 50 questions. If all 200 questions are honestly benchmarked it will be found, even in seemingly strong areas of performance, that weaknesses will be discovered. After all in every area of human endeavour, be it sport, family, or business, there will always be room for improvement. The Japanese have a philosophy of continuous improvement in life and their word for this is Kaizen (it is only in comparatively recent years that they have extended this word to include business endeavours).

Likewise it is not possible to limit the 200 questions to the factory. We accept that this book is aimed at manufacturing and that it is our argument that it is the adding of value in the factory that matters. But if you are a manufacturing company then surely everything else that is done in the organization is to support the endeavours of the factory? Or do you take the view that you are a marketing company and the factory is only there to answer the needs of the marketing function? After all it could be said that the marketing team is the interface with the customers, and therefore marketing knows what the customers want. Following this logic it would then be left to marketing to determine what shall be made (product), and because marketing 'know' what the market will pay (price) they would also determine the cost limits within which the factory must perform. Marketing would also have a fair say in how the product should be distributed (place) and, of course, how it should be advertised (promotion).

There are several dangers in this approach. David Packard (of Hewlett Packard) once said 'Marketing is too important to be left to the marketing department'. Kotler and Armstrong (1989) add that 'in a great marketing organization, you can't tell who's in the marketing department'. We don't totally disagree with these sentiments, but rather we would rephrase the statement to read 'in a world-class manufacturing company you can't tell who's in manufacturing and who's in marketing'. It is imperative that marketing and the manufacturing work as one. There is no room for functional divisions and jealousies.

Working together also means that, in the development of new products, marketing and the factory must be closely associated. Consider Question 30, 'How good and how frequent is your evaluation of your manufacturing capability to meet the impact of innovation?' At first glance it would appear that such a question should be directed to the factory, but on further consideration, perhaps this question should equally be directed to the marketing team.

In Chapter 3 we also discuss the need for full participation of manufacturing in product development. For some companies it would go without saying that new products are developed around the existing strengths and capabilities of the factory. But for many organizations the factory is ignored when the specifications and design of a new product are being determined. The factory will then be given a difficult product to make that requires retooling and retraining, with delivery expected within an impossible time frame. Put like that it sounds silly doesn't it? But it does happen! Recently an impressive TV advertising campaign launched a new product, but when we went to buy the item advertised it was not in the shops, simply because the factory had been unable to perform on time! In this case the advertiser created a demand and made sales for the opposition.

But total manufacturing solutions goes beyond working together as a team within the organization. In our supply-chain approach (see Chapter 4), we say that suppliers and customers should also be taken into our confidence and should become an integral part of the manufacturing process. While it is true that normally a customer only judges us on three occasions – that is when placing an order, when the goods are received, and when they get the account to pay (what Carlzon calls moments of truth) – we believe that a world-class manufacturer would want to get closer to the customer than this. We see no reason why customers should not be invited to give input to new product development, and we believe that customers should be encouraged to visit the factory at any time. Likewise with suppliers.

Many companies have little loyalty for their suppliers. If the aim is to get the best price and to use buying power to squeeze suppliers, how can a company expect suppliers to be anything but a little suspicious of the company? But if suppliers are treated not as adversaries, but as part of the planning team, then there are real benefits to be gained for the company and for the supplier. Suppliers can be invaluable in making suggestions for new product design, new materials and methods. Through their market intelligence, from a different perspective from your market intelligence, they can add to your knowledge of what the competitors are up to.

Consider the following scenario based on the old fashioned way of manufacturing. Our marketing team, against stiff opposition, have gained from a new client a large order. To get the sale they had to reduce our normal price. Because marketing had not fully understood manufacturing capabilities the product is going to be difficult to make. Delivery dates don't give us much leeway. Because of the fine margin, in placing our order for raw materials, price is considered to be the key factor in choosing a supplier. However we are careful to specify the standard, quantity and delivery dates. Materials are ordered to be delivered well in advance of when we need them, as we are not certain how reliable the supplier will be. In our order quantity, based on past experience, we also build in a factor for wastage during manufacture. When the materials arrive, somewhat late but not drastically so, we check them for quantity and quality. We find that we have been short supplied, and a large proportion is below our specified quality, also on further checking some material is found to be totally the wrong colour. We reluctantly keep and use the below standard material. We return the wrong materials and request urgent replacement and at the same time we also point out that we have been short supplied. The supplier is slow to reply, and the answer is unsatisfactory. In desperation we go to another supplier, this supplier is not sympathetic to our plight (they had missed out on our initial order because of price). They refuse to allow the discount we normally expect. They do however deliver on time and the material is up to standard. The same week that the second (expensive) supplier delivers, the original supplier makes good the shortfall. We now have more material than we really need, and neither supplier is prepared to accept a return of goods. The factory, as we had expected, wastes 10 per cent, and at the end of the run our quality inspector rejects 10 per cent of the finished goods. However, as we have fallen behind our delivery date and the customer is threatening to cancel the order, we deliver what we can including some of the below standard product to our client. Fifteen per cent of what we deliver is subsequently returned as being below standard, but by the time this happens we have been able to make and store sufficient replacements from our 'surplus' materials. We are therefore able to make good the order although we are now several weeks behind the original delivery date. Our customer takes 50 days to pay although our agreed terms were 30 days. Both our suppliers invoice us promptly. However there are errors on both sets of invoices, (wrong prices, in correct quantities, no credit for returns, and wrong discount calculations). We are obliged to pay most of the amount owing on the due date as we require further materials for a new order and neither supplier will supply until we have paid our outstanding accounts.

In the above scenario there were countless non-value-adding activities. These started with the problems of finding a supplier and of placing the order, checking the receipt of materials, returning materials, wasting materials in manufacture, manufacturing sub-standard goods, countless inspections and transportation activities, time wasted and wages expended on checking and arguing over invoices, and interest costs arising from cash flow problems (arising from having to pay for materials and the costs of manufacture before we receive payment).

Our first calculation shows that we broke even. But if we calculate the interest charge on the materials, and the time it took for us to be paid, and the overhead costs of ordering, checking and paying, plus all the problems associated with each stage of these activities, then our large order has cost us money. The unknown cost is the loss of our reputation with the customer who will be reluctant to buy from us again, and the loss of goodwill with both suppliers who now see us as difficult people to deal with.

What would be a better method? First marketing and the factory could work together when quoting for the order. If the customer was invited to the factory then maybe the new product could be designed around the manufacturing strengths of the factory. Certainly factory and client could establish a rapport with both understanding what is wanted and what is achievable. Possibly the customer will end up with a more suitable product. At the same time the preferred supplier, selected on reliability for quality and delivery time rather than on price alone, could also be invited to visit the factory and perhaps even meet with the client (now there is an interesting thought!).

Once the order is agreed, because we have faith in the supplier, only a sample of inward goods need be inspected. If each member of the factory knows what is expected, and can make their own tests against understood standards and because each operator is determined to get it right first time, then no quality inspections will be needed at the end of the line. If the customer is happy to receive goods in batches as they come off the line, then no finished goods will be held. And likewise wouldn't it be nice if our supplier delivered no more than the day before the materials were needed? Obviously our scheduling of work and materials will have to be good, and obviously our staff (purchasing, factory and accounts), will have to be efficient, know what is expected and know what corrective action they can take. For this scenario to work we have to be good at all levels, and we have to include our suppliers and customers as part of the team.

This is what we would call a total manufacturing solution.

Total manufacturing solutions also considers health and safety issues and includes being a good citizen of the universe. Gone are the days when it was acceptable to pollute your own rivers and environment: now it is no longer acceptable to transfer production off shore and pollute someone else's environment or to exploit their labour force. Apart from the hypocrisy of such an approach it will only be a matter of time before you are caught out. In the long run, if not for ethical reasons but certainly for good commercial reasons, it pays to do it right first time, rather than be embarrassed into making expensive corrections later. Chapter 5 discusses product safety, industrial safety and environment protection. The ethics of producing good 'clean' products and providing safe working conditions do not need elaboration. The economic sense of doing so is also self-evident. Chapter 5 considers some of the actions that need to be taken to protect your products. It covers what to look for when receiving materials, how to store materials and how to prevent contamination during processing. Safety, accident prevention and emergency procedures are also covered.

Chapter 6 goes into the details of manufacturing facilities. In this chapter we return to the theme of team work and the need for marketing and the factory to work together and for both functions to understand the capabilities and strengths of your operation. The question of flexible manufacturing is considered. While we accept that there are considerable advantages from being flexible, we do make the point that forward planning and design can reduce the need to be flexible all of the time. We do not suggest that flexible manufacturing is a bad thing, or that in all circumstances it will benefit a company, but we do believe that many of the benefits of just-in-time can be achieved with large batches. It does not necessarily follow that making everything in batches of one just to achieve flexibility is always the best option. The aim is not flexibility, or just-in-time, or single-minute exchange of dies; the aim is efficiency. Economies of scale should still be considered as a viable option to flexible manufacturing and in some circumstances batch production will prove to be the more efficient option. The real issue is manufacturing performance.

Manufacturing performance is dependent on many things, the skills and ingenuity of the factory workers, the efficiency of the supply of the materials, the scheduling of work, the equipment we have, and not least the reliability of the equipment. Quality includes the level of maintenance of the plant. We consider that maintenance is a competitive weapon. Question 129 asks 'How well is it understood throughout your organization that maintenance is a competitive weapon?'

To determine our manufacturing performance we are firm believers in measurement. The often maligned F.W. Taylor once said 'if it can't be measured it can't be managed'. We agree. And we say this applies to all aspects of a company, and thus we have 200 questions which enable you to obtain a benchmark measurement for the whole company and to extend them to include relations with suppliers and customers. With measurement of manufacturing performance we provide a series of quantifiable measurements which will give a precise measurement. Before any measurement should be attempted it is desirable that we know why we are measuring and what actions we can take as the result of the measurement. We say that the mere act of taking a measurement will mean time and effort, and that measurement in itself is not a value-adding activity. If measurements are unreliable, or if the results are of interest only, there is no benefit in taking them. To justify the time and cost of taking a measurement it is important that the results are studied and that some concrete action is taken. The aim is to improve the quality of what we are producing, 'quality' being the reduction of non-value-added activities and efficiency in our value-adding activities.

Much of what we already measure is in the form of accounting information, and much time and effort go into recording this information. We are of the opinion that often data is gathered just for the sake of gathering it, and little effort is made to constructively use the information we already have. We actually don't need more information, but we must make sure that the information we are receiving is relevant and that we are using it to our advantage. In Chapter 7 we look at various standard accounting ratios and discuss the importance and relevance of each. This chapter also explains why we need a strategy for information technology. We live in a age where information has never before been so readily available. The danger is that we can get swamped with it. We also have to be aware of the pitfalls of computerization and of some of the basic safeguards that have to be taken when considering a new system.

We have said earlier that quality is a cultural issue. Quality and the elimination of waste, the continuous search for better methods and processes should become so engrained into the psyche of the people of the organization that they are not even conscious that they are practising total quality management. Instinctively when things go wrong they will know what to do and always react in the way that management would hope that they would react. Before this state can be achieved much learning has occurred.

Chapter 8 is our people chapter. The study of organizational cultures is not the study of organizations as such but is the study of the people who make up the organization. The hackneyed expression 'people are our greatest resource' is in fact more than a cliché. Or in the words of a New Zealand Maori proverb:

> What is the important thing?
> It is people
> It is people
> It is people.

Japanese manufacturers are well aware of the power of the people. They think that manufacturing in the West is still bound by Taylorism where the bosses do the thinking and the workers wield the screw drivers. The Japanese, on the other hand, believe that they encourage all the workers to help with the thinking. In practice their systems are fairly rigid, and although suggestions are encouraged they tend to be group suggestions made by teams such as quality circles rather than suggestions made by individuals. Before any suggestion is adopted it goes through a rigorous examination. If implemented it then becomes documented as standard practice. No changes to the standard practice are permitted without an examination using a set procedure. With Taylor it was the bosses' job to find the best method and then to enforce it. With the Japanese approach the workers themselves find the best method and enforce it on themselves. That is the Japanese approach to empowerment.

In Chapter 8 we discuss our approach to learning and empowerment. We believe the key is through education and the provision of learning opportunities for everyone in the organization.

In Chapter 9 by asking the 200 questions, we provide the basis for the answer of how total manufacturing solutions might be achieved. But, before the 200 questions can be asked, a philosophy and a method are needed.

The method includes forming a team to ask the questions, and a system for the capture of data. These issues are covered in Chapter 10. Suffice it to say that as total manufacturing solutions embraces the whole company, it follows as surely as night follows day, that the team must be drawn from each major function of the organization. Much of the data on which to make the answers to the questions will already exist in the company. In Chapter 10 we discuss where data can be found.

Chapter 11 considers the philosophy of excellence and the magnitude of change needed to establish a culture of quality throughout a company. Cultural changes cannot be limited to just one section of an organization, any attempt to do so will be doomed to failure. The whole company, from the chief executive to the cleaner, must share the same vision and have the same goals. Likewise it is not possible for any one function to work in isolation or to remain aloof. Although it is important that direction and leadership come from the top of the company, unless everyone at all levels shares the same passion for manufacturing correctness and excellence, it won't happen. The culture has to be all pervasive. Changing a culture does not happen overnight. But unless the decision to change is made and steps are taken to effect a change it will never happen. Changing a culture will not be an easy task.

Once the culture is right (the philosophy has been explained and accepted, i.e. everyone is living and breathing the philosophy), and the team (drawn from senior experienced people from each of the key functions) has been appointed and trained, and the data has been collected, then (and only then) can an attempt be made to seek answers to the 200 questions.

Initially the 200 questions will not provide a solution. The 200 questions have been designed in such away that the initial aim is to score or rate the company against each question. There is no short cut. Each question has to be considered and a rating agreed. In effect you will be using the questions as a means of benchmarking your company. Having calculated a benchmark score the next stage is to analyse the results. The purpose of the analysis being to establish where the weaknesses are and to consider what actions are necessary to improve the situation. The gap between where your company is and where it needs to be will be obvious once the rating has been completed. Each company will have their own pattern of gaps. It is unlikely that any two companies will have the same pattern. An indication of what various gaps mean is given in Chapter 12. Knowing your benchmark position and having determined what is required facilitates the establishment of an improvement strategy.

Improvement strategies are discussed in Chapter 13, and the management of the improvement process, plus the place of consultants in the improvement process, is considered in Chapter 14. However, the overall emphasis is on self-examination to enable a company to benchmark against world-class standards. The beauty of our approach is in its simplicity. You do not have to enter into a partnership with another organization and trade information to see how well you compare to their performance. With our approach, by gathering data for the 200 questions, and then by self-examination of your performance for each of the questions, you can establish your manufacturing correctness factor against our established world-class standards. We give you the means to establish an external benchmark position without having to go outside the company!

From this examination we show which of five categories of company you currently fall into and what this means. An important issue in determining which category is the relationship and relative values of performance and practice issues. We explain how performance cannot be sustained if the supporting performance pillars are weak. For each category we highlight actions that might be taken so as to achieve (eventually) world class status.

Chapter 14 explains the implementation process. We make the point that many organizations have gone through the agony of benchmarking, or have gone to lengths to determine strengths and weaknesses, but if nothing happens all the effort (time and money) will be for nothing. Implementation cannot be rushed into. Our approach is a structured approach centred on a project team. We stress the need for an 'open' programme well publicized throughout the company with no hidden agendas. Achieving total manufacturing solutions could well take in excess of five years and even then perfection will not be achieved. Ours is an ongoing programme.

In conclusion we restate our philosophy which is that it is both the achievement of manufacturing performance and the application of best practices that give a company a sustainable leading edge. Further we believe that history proves that strong nations are those who are strong in manufacturing. The Industrial Revolution made Britain Great, mass production and innovation made the United States a world power, and manufacturing with a focus on quality has made Japan the strong industrial nation it now is. To sustain a strong healthy national economy a country needs efficient world class manufacturers. Today's issues are globalization and international competition. No longer are companies protected by national barriers and tariffs. For some this may be seen as a threat, but conversely, an efficient organization will see this as an opportunity. This book shows how to make the most of the emerging opportunities.

References

Kotler, P. and Armstrong, G. (1989) *Principles of Marketing*. Prentice Hall.

'ASK' – Self Assessment Program: Benchmarking made easy

The 'ASK' software, which is complementary to the book, provides a truly practical way of monitoring and improving business performance. This invaluable and user-friendly tool may be used by managers, individuals or class based groups as part of regular training.

Incorporating 200 questions relating directly to the six 'pillars' of Total Manufacturing and the twenty associated 'foundation stones', the user can evaluate how well their own organization is performing and identify any necessary remedial steps. Assessments are scored and a series of visual graphs used to identify positioning in comparison to the ideal and to previous assessments. The user can control the setting of target scores and/or weighting for each of twenty 'foundation stones' as appropriate.

The 'ASK' software for Windows™ is available from The Hampton Technology Group, Westfield House, Hampton Court Road, Hampton Court, Surrey KT8 9BX (order line tel: 0181 977 1699) (diskette + manual).

Index

Accidents, 57, 58, 62, 66
   cause, 64–5
   control, 65
   cost, 63
   HAZOP, 65
   prevention, 65
   ratio triangle, 62
Accounting, *see* Financial management
Acronyms, 1–8, 253
Activity based costing, 47, 100, 133
Analysis: Computer aided, 200–1
Appropriate technology, 73, 80, 81, 84, 87
Article numbering, 40
Assembly process/line, 11, 159
   alternative Volvo experiment, 159
Asset productivity, 86
Automation, 138

Baldrige Award, 118
   see also Quality and TQM
Batch production, 10, 11, 39
Behaviourists, 98, 203
Benchmarking, 2, 14, 26, 80–1, 99, 102, 171, 173, 189–90, 193, 204, 217, 222, 224, 226, 229, 235–8, 242–3, 248, 251, 253, 258, 260–1
Best practice, 2, 4, 66, 143, 180, 192–4, 199, 204, 215, 225–31, 242, 247
Big picture, 13, 252–61
Bills of Material, 36–9
Boston Consulting Group, 23–4

Brand image, 20, 22
Bureaucratic management/organizations, 115, 144
Business plan, 36, 75, 174
Business process re–engineering (BPR), 33, 140–1, 148, 162, 166, 221, 227, 229, 232
Business process, 29, 33, 49, 90, 140–1,191, 201, 222, 225, 233
   see also Total process

CAD/CAM, 84, 136
Cause and effect, 224–5
Change management, 145, 165–6, 208–10, 238, 243–6, 259
Channels of distribution, 42
Communication, 36, 245–6
Competition/competitors, 1, 22, 26–8, 173
Computer aided analysis, 200–1
Computer integrated manufacturing, 84
Computers and systems, *see* Information Technology
Contribution, 22
Core business/strengths, 22, 28, 229–30
Costs and cost cutting, 134–5, 256
Critical success factors, 58
Culture, 33, 92, 98, 189
   Quality and TQM, 113, 116, 144–5, 206–9
Customer:
   composite customer service index, 55
   cycle time, 52
   distribution, 40, 175–6

ISO 9000, 118
partnerships, 39, 46
potential customers, 21
profitability, 48
quality perceptions, 110
real customers, 21
requirements, 17, 20
service, 54–5
TQM, 114
Cycle time, 52

Data
  analysis, 77, 189–204
  collection, 77
  exchange, 40
  manufacturing, 77
  preparation, 190
  recording, 6, 189–204, 242
Delivery performance, 53
Denominator management, 85–6
Design:
  warehouse, 43
Discounted cash flows, 87, 133
Distribution, 40–6
  channels, 42
  customers, 40
  fast moving consumer goods, 41
  management, 40, 45, 174–5
  performance, 53
  planning, 37
  requirements, 34
  routes, 46
  strategy, 41
  third party, 41
  transportation, 45

Electronic data interchange, 40, 136
Electronic inventory, 45
Empowerment/team work, 115, 141, 161, 259
Energy efficiency, 101
Environment, 27
  environment and safety, 57–69
  environment protection, 66–7
  questions, 177–9
Ethics, 57, 257
European article numbering, 40

Fast moving consumer goods (FMCGs), 12, 41, 51
Faulty products, 58
Final analysis, 253
Financial management, 125–35, 184
  cost cutting, 134
  key concepts/definitions, 126–7
  objectives, 126
  ratios, 127–32
  strategic costs factors, 132
  systems, 133
Fire fighting:
  equipment, systems and procedures, 65
Five whys, 10
Flexible manufacturing and work practices, 84, 88–91, 186–7, 257
Ford, Henry, 88
Foundation stones, 3–5, 9, 12, 171, 215, 253
Functional boundaries, 34–5, 144
  cross function teams, 145

Gap analysis, 6, 28, 215–26, 242–3
  Normalizing procedures, 217–18
  tools, 218–21
  setting targets, 215–16
General Motors, 136
Global market/globalization, 1, 68, 74, 252
Great Britain and The Industrial Revolution, 1, 261
Green Movement, *see* Environment

Hazardous Operations (HAZOP),

65
Henley, 149
Human Resources, *see* People

Implementation, 235–51
  feedback, 247
  planning, 243–4
Improvement process, 7
  strategy, 227–34
Incremental improvement, 33, 253
  see also Quality and TQM
Industrial engineering, 160
Industrial relations, 160–2
Industrial Revolution, 1, 261
Industrial safety, 61–6, 178
Information flow, 34–5
Information Technology, 135–41, 185
  application software, 139
  General Motors, 136
  hardware, 137
  implementation, 140–1
  incompatible systems, 136
  open systems, 137
  project teams, 140–1
  rapid growth of, 135–6
  software, 136, 138
  standards, 137–8
  strategy, 136–8
Innovation and marketing, 4, 17–19
Innovation and process, 28–32, 173–4
Installation, *see* Implementation
Integrated innovation, 30
Inventory, 44–5
  see also Stock
ISO 9000 Series, 120–5

Japan:
  five whys, 10
  investment capital & workers, 85
  Kaizen, 253

lean production, 152
operations management, 90
SMED, 90
strength and practices, 1–3, 261
Toyota, 11, 89
TPM, 94

Labour, 74
  productivity, 100
  see also People and Culture
Laggers, 203, 233
Leaders (world class), 203, 230
Leadership, 151–2
  see also Management
Learning, 152–70, 187–8
  change process, 245
  continuous learning, 162–4
  crafts flexibility, 153
  for change, 165–6
  for new technology, 164–5
  multi skilling, 157
  performance, 168–9
  resources, 167
  skills flexibility, 153–5
  task sharing, 155–6
Logistics, 34, 40
  see also Supply chain

Maintenance, 92–8
  breakdown, 93
  competitive weapon, 98, 182, 257
  condition based predictive, 93
  infrastructure, 94
  measurement, 93
  mix, 95
  optimum, 93–4
  performance, 98
  policy, 94
  predictive, 93
  reliability centred maintenance, 94–5
  reliable manufacturing, 92
  time based, 93

Total Productive Maintenance (TPM), 94–7, 152
  trends, 92
Management, 147–52, 185–6
  leadership, 151–2
  obstacles, 70, 149–51
  profile, 147–8
  training, 148, 149
  see also Teamwork and people
Manufacturing, 70–7, 88–98, 257–8
  advantages, 99–100
  competitive weapon, 2
  computer aided analysis, 200
  computer integrated, 84
  data collection, 77
  definition of, 3
  effectiveness, 104–5
  efficiency, 100
  facilities, 70
  feasibility, 18
  flexibility, 88–91, 106, 181, 257
  globalization, 74
  low labour costs, 74
  manufacturing correctness factor, 6, 8, 16, 198–200
  marketing, 18–19
  missing link, 57
  mission, 75
  performance, 73, 182–3, 257–8
  planning, 73
  process map, 223
  processes, 10–11
  reliable, 73, 92–8, 182
  resource planning, 36–7, 40, 174–5
  set up time (reduction), 87
  SMED, 90, 257
  sourcing, 74, 179–80
  strategic role, 71
  strategy and factors, 72, 76, 179–80
  technology, 180–1
  see also TMS
Manufacturing resource planning (MRP II), 36–7, 40, 174–5
Marketplace/share 172, 252, 254
  see also Globalization
Marketing, 4, 17–32
  and manufacturing, 18–19, 254
  see also Customers
Materials:
  flows, 34–5
  productivity, 100–1
  see also Manufacturing resource planning, and Supply chain
Measurement, 20–3, 258
  importance of, 98–9
  learning performance, 168
  operations performance, 52
  safety, 60, 66
  supply chain, 49–50
  Taylor, F. W., 99–107, 258
Meetings, 36
Mission statements, 14–15, 205–14
  gap analysis, 6–7
  making it happen, 15–16
  manufacturing, 75
  wording of and writing, 14–15
Moments of truth, 254

Net Present Value (NPV), 87

Obsolescence, 40
On time delivery, 53
Operating time, 102
Operations performance, 52
Operations plan, 36
Order cycle time, 52
Order flow, 37
Organizations/Structures, 10, 12, 17, 115, 144–7, 185–6
  borderless, 147
  leadership, 151
  matrix structure, 146
  obstacles, 149
  paradigm, 150
  paradox, 149
  perplexity, 150

power, 149
Outbound logistics, 40

Plant efficiency and capacity
  evaluation (PEACE), 102–3
Pareto (80/20 Rule), 47, 222
Partnerships:
  customers, 46
  suppliers, 174
People, 115, 141, 143–69, 257–9
  flexible working practices,
    152–7
  industrial relations, 160–2
  leadership, 151
  management profile, 147
  management skills and culture,
    144–5
  maori proverb, 259
  team structuring, 157–60
  training, 148, 162–9
Performance improvement, 105–6
Performance Vs Practice, 201–2
Pillars, 3–5, 70, 253
  environment and safety, 57–69
  manufacturing facilities, 70–108
  marketing and innovation,
    17–32
  people, 147–70
  procedures, 109–46
  supply chain management,
    33–56
  supply chain, 9, 11, 13
Pipeline map, 51–2
Plan, plans, planning
  manufacturing facilities, 106
  manufacturing resource
    planning, 36–7, 174–5
  marketing, 20
  supply chain, 34–7
Plant efficiency, 102–4
Plodders, 203, 232
Pollution, *see* Environment
Post delivery, 53
Priorities, 12, 37

Problems/priorities/total
  approach, 10, 12–14
  and quality, 117–19
  critical analysis, 219
  five whys, 10
Procedures, 109–42
  financial management/
  Information Technology/
  Quality Management, 109–25
Process innovation, 28–31, 173–4
  design, 244–5
  re–engineering, 229
Procurement:
  push and pull, 44
Product:
  customer requirements, 17, 254
  definition, attributes etc., 17,
    254
  development strategy, 28, 254
  integrated product line management, 29–30
  life cycle, 29–31
  portfolios, 22
  process innovation, 28–31,
    173–4
Product safety, 58–61, 176
Production process:
  job, batch, assembly, 10–11
  schedule, 44
Project/s, 75
  implementation project, 238–45
  project brief (manufacturing),
    75
  project manager/leader, 239

Quality, 1, 13, 20, 33, 109–25,
  183–4, 253–61
  and workers, 107
  assurance, 112
  circles, 259
  control, 112
  culture, 113, 258
  customers, 114
  definition, 109

empowerment, 141
hierarchy, 110
inspection, 111
ISO 9000 series, 120–5
problem solving tools, 117–19
statistic process control, 113
suppliers, 114
Total Quality Management, 113–17, 125
world class status, 253, 260

Re-engineering, 140, 166
Reliable manufacturing, 92
  see also Maintenance
Re-order level, 44
Re-order quantity, 44
Research and development, 28
Return on investment (ROI), 85
Robotics, 80

Safety, 58–65, 257
  hazards, 59
  HAZOP, 65
  industrial, 61–3, 178
  manager, 66
  measures, 60, 66
  product, 58–61, 177
Scheduling, 34
Self analysis, 27, 242–3
Service (competitive weapon), 1
Service levels, 21
Seventy-two hour car, 11, 89
  see also Toyota
Single minute exchange of dies (SMED), 90–1, 257
Six big losses, 98
Spider diagram, 198–9
Stock, 38–9, 44–5
  control, 34
  slippage, 40
Stock keeping units (SKUs), 38–9
  management, 44–45
  profile, 51
  re-order systems, 44–5

work in progress, 45
Strategic, Strategy, 36, 57, 78–9, 80–7
  cost factors, 132
  improvement, 227–34
  Information Technology, 136–40
  manufacturing facilities, 106
  sourcing strategy, 71–80
Structuralists, 203, 231
Supply chain, 9, 11, 33–56, 191–2, 224, 254–6
  electronic data interchange, 136
  Information Technology, 138
  logistics, 34, 40
  management, 33–56
  performance, 48–9, 176
  suppliers + TQM, 114
  suppliers, 35–6, 39
  working with the suppliers, 174, 254–6
SWOT Analysis, 27

Teams/team work, 98, 115, 141, 150, 254
  benchmarking team, 190–1
  industrial engineering, 160
  Taylorism, 158
  team structuring, 157–60
  TIED analysis, 159
  Volvo, 158–60
  see also People and Culture
Technology/technological, 22, 80–7
  advances, 27
  appropriate technology, 80–7, 180–1
  choice, 84
  evaluation, 85, 87
  model, 81–2
  strategy, 80–7
Total business process, 71
Total Manufacturing Solutions (TMS), 3–6, 9–16, 215, 252–61
  definition, 3

foundation stones, 4
pillars, 4
process, 7
self assessment, 5–6
two hundred questions, 9–10, 171–88
Total process, 12, 42, 94, 147, 149, 162, 191
Total Productive Maintenance (TPM), 93, 98
Total Quality Management (TQM), *see* Quality
Toyota, 11, 89
Training, 148, 162–9, 245–6
see also Learning
Transformation process, 9, 159
Transportation, 41, 45
Two hundred questions, 9–10, 171–88, 191–6
  assessment and scoring, 196–204
  the competition, 173
  customers, 175–6
  distribution management, 175
  environment protection, 178–9
  financial management, 184
  flexible manufacturing systems, 182
  flexible working practices, 186–7
  industrial safety, 178
  Information Technology, 185
  learning, 187–8
  management skills, 185–6
  manufacturing performance, 182–3
  manufacturing resource planning, 174
  manufacturing technology, 180
  the market place, 172
  organization, 186
  product and process, 173–4
  product safety, 177
  quality management, 183–4
  reliable manufacturing, 182
  sourcing strategy, 179–80
  suppliers, 174–5
  supply chain, 176
  why, 9–10, 171–88

Value analysis, 28, 135
Value chain, 191–3
  see also Supply chain
Vision, 207–9

Warehouse/warehousing, 34, 40, 42–3
World class manufacturing, 73–4, 230–1, 260
  contenders, 74
  leaders, 230–1

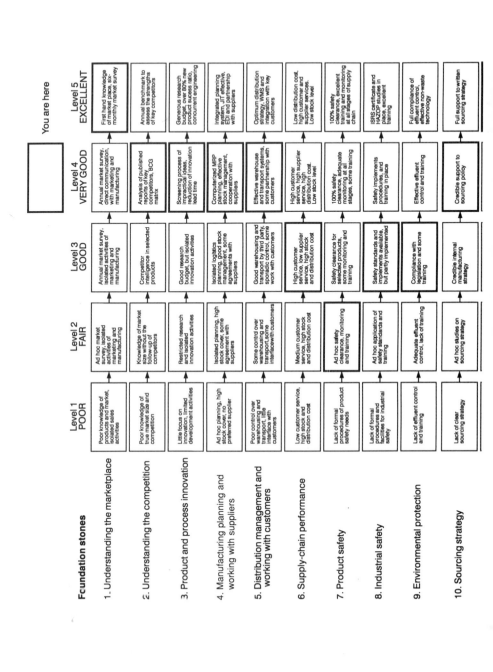

Criterion	Level 1	Level 2	Level 3	Level 4	Level 5
11. Appropriate manufacturing technology	Lack of a clear policy on technology	Ad hoc appraisal of technology and investment	Capital investment primarily based on ROI	Credible support to policy on technology and investment	Full support to written policy on technology and investment
12. Flexible manufacturing	Lack of understanding of flexible manufacturing	Some applications of flexible manufacturing	Good understanding, partial application of flexible manufacturing	Credible support to flexible manufacturing	Full support to flexible manufacturing and quick change-over
13. Reliable manufacturing	Breakdown maintenance, low reliability, high cost	Time based maintenance, high maintenance cost	Condition based maintenance, some maintenance control system	Partial TPM, high reliability, high maintenance costs	Full implementation of TPM, high reliability, low maintenance cost
14. Manufacturing performance	Low level of manufacturing performance, poor reporting system	Medium level manufacturing performance for some plants	Medium level manufacturing performance, limited IE support	Sporadically high manufacturing performance with limited IE Support	High manufacturing performance with integrated industrial engineering support
15. Quality management	Lack of defined procedures and standards for quality	Quality management in selected areas without top management support	Some use of SPC tools and quality procedures	Use of SPC tools and customer/supplier culture	Baldrige award or ISO9000 winner, change culture
16. Financial management	Lack of defined financial management, poor ROI	Medium ROI, and cash flow, no cost effectiveness	Sporadically good ROI and cash flow, some cost effectiveness	Good ROI and cash flow and cost effectiveness	Excellent ROI, cash flow and cost effectiveness programmes
17. Information technology	Lack of IT strategy, support and application in manufacturing	Some good application of hardware and software in manufacturing	Customized application software in all areas of business	Open system IT, strategy not implemented, integrated network	Open system IT strategy in place, integrated client-server network
18. Management skills and organization	Lack of leadership, organization and management development	Traditional layers of management, some good managers	Good management team without development plans	Good management team and creditable management development plan	Visionary leadership, and policy to attract and develop high calibre managers
19. Flexible working practices	Restricted working practices, poor industrial relations	Traditional working practices, good industrial relations	Limited flexible working practices and some consultation with unions	Multiskilled workforce, limited consultation with unions	Effective partnership with unions and sustainable working practices
20. Continuous learning	Lack of learning resources and training plans	Selective training plans by external resources	Limited learning resources and selective training plans	Good learning resources and selective training plans	Learning organization with written training plans for all employees